T0275857

Integrating Green and Sustainable Chemistry Principles into Education

Integrating Green and Sustainable Chemistry Principles into Education

Edited by

Andrew P. Dicks

Professor, Teaching Stream
Department of Chemistry, University of Toronto
Canada

Loyd D. Bastin

Professor
Departments of Chemistry and Biochemistry
Widener University
USA

ELSEVIER

Elsevier
Radarweg 29, PO Box 211, 1000 AE Amsterdam, Netherlands
The Boulevard, Langford Lane, Kidlington, Oxford OX5 1GB, United Kingdom
50 Hampshire Street, 5th Floor, Cambridge, MA 02139, United States

Integrating Green and Sustainable Chemistry Principles into Education

Copyright © 2019 Elsevier Inc. All rights reserved.

No part of this publication may be reproduced or transmitted in any form or by any means, electronic or mechanical, including photocopying, recording, or any information storage and retrieval system, without permission in writing from the publisher. Details on how to seek permission, further information about the Publisher's permissions policies and our arrangements with organizations such as the Copyright Clearance Center and the Copyright Licensing Agency, can be found at our website: www.elsevier.com/permissions.

This book and the individual contributions contained in it are protected under copyright by the Publisher (other than as may be noted herein).

Notices
Practitioners and researchers must always rely on their own experience and knowledge in evaluating and using any information, methods, compounds or experiments described herein. Because of rapid advances in the medical sciences, in particular, independent verification of diagnoses and drug dosages should be made. To the fullest extent of the law, no responsibility is assumed by Elsevier, authors, editors or contributors for any injury and/or damage to persons or property as a matter of products liability, negligence or otherwise, or from any use or operation of any methods, products, instructions, or ideas contained in the material herein.

ISBN: 978-0-12-817418-0

Publisher: Susan Dennis
Acquisition Editor: Anneka Hess
Editorial Project Manager: Emerald Li
Production Project Manager: Kiruthika Govindaraju
Cover Designer: Matthew Limbert

Contents

CHAPTER 6   Promoting political and civic engagement in a nonmajor sustainable chemistry course 141

Loyd D. Bastin, PhD, Andrea E. Martin, PhD

CHAPTER 7   Development and implementation of a bachelor of science degree in green chemistry 163

Nicholas B. Kingsley, PhD, Jessica L. Tischler, PhD

Contributors

Loyd D. Bastin, PhD
Professor, Departments of Chemistry and Biochemistry, Widener University, One University Place, Chester, PA, United States

Rachel M. Chapman
Undergraduate Chemistry Major, Department of Chemistry, Radford University, Radford, VA, United States

Andrew P. Dicks, PhD
Professor, Teaching Stream, Department of Chemistry, University of Toronto, Toronto, Ontario, Canada

Thomas Holme, PhD
Morrill Professor, Department of Chemistry, Iowa State University, Ames, IA, United States

Glenn A. Hurst, PhD
Assistant Professor, Green Chemistry Centre of Excellence, Department of Chemistry, University of York, York, United Kingdom

Sarah A. Kennedy, PhD
Assistant Professor, Department of Chemistry, Radford University, Radford, VA, United States

Nicholas B. Kingsley, PhD
Associate Professor of Chemistry, Department of Chemistry and Biochemistry, University of Michigan-Flint, Flint, MI, United States

Renuka Manchanayakage, PhD
Assistant Professor, Department of Chemistry, St. John Fisher College, Rochester, NY, United States

Andrea E. Martin, PhD
Associate Professor, Department of Chemistry, Widener University, One University Place, Chester, PA, United States

Manisha Nigam, PhD
Associate Professor, Department of Chemistry, University of Pittsburgh at Johnstown, Johnstown, PA, United States

Jessica L. Tischler, PhD
Associate Professor of Chemistry, Department of Chemistry and Biochemistry, University of Michigan-Flint, Flint, MI, United States

Penny S. Workman, PhD
Associate Professor, Department of Chemistry, University of Wisconsin — Stevens Point at Wausau, Wausau, WI, United States

Coeditor biographies

Loyd D. Bastin is currently a professor of Chemistry and Biochemistry and coordinator of undergraduate research at Widener University where he has taught organic chemistry, biochemistry, and sustainability since 2004. Since 2000, he has been interested in the development of green chemistry and sustainability materials for undergraduate chemistry curricula. He has authored several peer-reviewed articles and book chapters discussing the incorporation of sustainability, environmental justice, and green chemistry into the curriculum. He has organized green chemistry and sustainability-related symposia at the past five ACS Biennial Conferences on Chemical Education. His current area of research is the development of greener methods for synthesizing pharmaceuticals and the development of new laboratory experiments for the undergraduate chemistry curriculum. He currently serves as cochair of the Widener University Sustainability Council and is a member of the Chester Environmental Partnership.

Andrew P. Dicks holds the position of Professor, Teaching Stream at the University of Toronto where he arrived as a postdoctoral research fellow in 1997. He has research interests in undergraduate laboratory instruction that involve designing novel and stimulating experiments, particularly those that showcase green chemistry principles. He has edited two books as resources for teaching green chemistry (*Green Organic Chemistry in Lecture and Laboratory* and *Problem-Solving Exercises in Green and Sustainable Chemistry*). He is the recipient of several pedagogical awards, including the 2011 American Chemical Society-Committee on Environmental Improvement Award for Incorporating Sustainability into Chemistry Education and the 2015 Chemical Institute of Canada National Green Chemistry and Engineering Award (Individual Category). In 2014, he was cochair of the 23rd IUPAC International Conference on Chemistry Education which was held in Toronto.

Foreword

The Earth needs scientists who can speak for it.
Jennifer Ritzmann

Few of us would argue that education is the key to developing outstanding scientists. But growing scientists that care for the Earth requires more—much more. It requires educators who believe deeply in sustainability. It requires a community of these educators to share their knowledge and experiences with like-minded scholars. Lastly, it requires that this community embark upon a journey of dedicated teaching, learning, and research based on their shared knowledge, experience, and creativity. This is not something new to green chemistry. For well over a decade now, symposia, posters, and informal gatherings at the Biennial Conference on Chemical Education (BCCE) have provided the information as well as the impetus essential to growing a community of exceptional green and sustainable chemistry scholars.

During the summer of 2018, 25th BCCE conferees at the University of Notre Dame attended 31 presentations in two green chemistry symposia: "To Green or Not to Green? Approaches for Including Green Chemistry in a Traditional Academic Setting: Teaching, Research & Service" organized by Amy Cannon and Glenn Hurst; and "Green Chemistry in High School, College & University Curricula (& Beyond!)" organized by Loyd Bastin and Andrew Dicks. The latter symposium was divided into four sessions: green chemistry in the organic laboratory; green chemistry courses; integration of green chemistry and sustainability; and green chemistry in international high schools. Several other presentations on the theme of green chemistry education were offered in other symposia including "Engaging Students in Physical Chemistry" and "Chemistry Education Research: Graduate Student Research Symposium."

This book, drawn from presentations at the 2018 BCCE, reflects the multiple drivers of the green chemistry movement. While much of the movement was initiated in organic chemistry with industrial applications, the chapters also reflect the more current emphases on teaching strategies, student research, diversity of chemistry disciplines, community-based learning, and program development.

The opening chapter by Sarah Kennedy and Rachel Chapman illustrates current teaching methods that focus on inclusive and impactful pedagogy, provides examples of green chemistry courses, and lists an extensive discussion of book, journal, and institutional resources. Thomas Holme's chapter uses the frameworks of planetary boundaries and systems thinking to infuse sustainability into the introductory chemistry curriculum.

Penny Workman stresses the importance of undergraduate research and illustrates how green chemistry can be the central theme of student research in organic chemistry. Undergraduate research is also the topic of Andrew Dicks who investigates the efficacy of using microwave energy in organic teaching laboratories, describes contemporary microwave research, and challenges his students to compare the energy efficiency of a microwave-assisted versus a conventionally heated reaction.

Following this, Renuka Manchanayakage demonstrates how green chemistry can be successfully applied to service-learning in the content areas of water quality and renewable energy. Similarly, Loyd Bastin and Andrea Martin embed civic engagement into a nonmajors chemistry course through social justice advocacy, field experiences, and investigations into legislative actions based on the theme of chemistry as the "central science."

At an institutional level, the lack of incorporation of green chemistry into standard curricula prompted Jessica Tischler and Nicholas Kingsley to develop and report on the challenges and successes of devising a new green chemistry major program. Manisha Nigam describes a broader vision of sustainability both throughout a university and within an advanced-level interdisciplinary course. Finally, Glenn Hurst closes the volume by presenting work toward creating a portfolio of transferable resources to teach green chemistry internationally in developed and developing countries and for multiple levels of students.

The green chemistry symposia at the BCCE will continue to be a core that grows scholars who can speak for the Earth. Throughout this book we can see the exceptional steps the authors have taken to move their scholarship toward a chemistry education that is greener and more sustainable. Certainly, the strong presence of them and many more scientists who share the ideals of a sustainable Earth will continue the tradition at the 26th BCCE at Oregon State University in 2020!

Professor Susan H. Sutheimer
Green Mountain College
March 2019

Acknowledgments

We would like to thank Susan Sutheimer for her inspirational foreword to this book and for her support and guidance in organizing green chemistry education symposia at several BCCE conferences during the last decade. Amy Cannon and Glenn Hurst were instrumental at the 2018 BCCE in organizing the "To Green or Not to Green? Approaches for Including Green Chemistry in a Traditional Academic Setting: Teaching, Research & Service" symposium from which several chapters in this book are drawn. Finally, the publishing team at Elsevier (Anneka Hess, Alexandra Romano, and Jennifer Horigan) is also gratefully acknowledged.

Loyd D. Bastin
Andrew P. Dicks
March 2019

Green chemistry as the inspiration for impactful and inclusive teaching strategies

1

Sarah A. Kennedy, PhD [1], Rachel M. Chapman[2]

Assistant Professor, Department of Chemistry, Radford University, Radford, VA, United States[1];
Undergraduate Chemistry Major, Department of Chemistry, Radford University, Radford, VA,
United States[2]

1.1 Introduction

Twenty years ago, the 12 Principles of Green Chemistry outlined by Paul Anastas and John Warner created a clear call to chemists and educators to rethink how we practice chemistry, and the responsibility we have as scientists and educators to the sustainability of our world *(1)*. Today, these 12 Principles still provide a guide for making changes in how we operate as chemists, and lay the foundation for decisions made in education, industry, and research laboratories. Green chemistry is being infused into higher education curricula as seen by the growing number of articles in the *Journal of Chemical Education* and other academic publications. Approaches by green chemistry educators have varied from changing one experiment in a course to completely revamping an entire laboratory curriculum. Many individual experiments (especially in the organic laboratory curriculum) have been developed for adoption. Morra and Dicks recently discussed a collection of these experiments and organized them by technique and pedagogy *(2)*. Other individuals have created complete courses focused on green chemistry *(3–8)*.

Sustainability and green chemistry are areas of study that naturally excite students and provide an avenue for engaging future scientists and nonscientists in a deep and meaningful way. In addition to the course content being relevant, timely, and necessary, the green chemistry classroom can provide an effective environment by incorporating high-impact practices (HIPs) and inclusive pedagogy strategies that engage students in learning. These approaches provide a welcoming and engaging environment for *all* learners by considering and embracing their differences. Green chemistry courses have the potential to provide the most valuable learning experiences that an undergraduate may have, based on the crucial content and memorable way that each student can engage with the course concepts.

This chapter begins by providing some background about research-based inclusive and impactful pedagogical practices that are recognized by national organizations focused on higher education. These practices can be introduced into the green chemistry classroom, and align with the American Chemical Society's (ACS) Committee on Professional Training (CPT) *(9)*. This is followed by a review of several green chemistry courses with a focus on their use of pedagogical strategies. A case study of one class designed and taught by the author showcases ways that teaching practices and a focus on inclusivity can be utilized to teach an upper-level green chemistry class. Finally, a substantial collection of green chemistry educational materials is presented to highlight resources to educators that plan to incorporate green chemistry into their courses.

1.2 Effective pedagogy and green chemistry education

Traditionally, chemistry has been taught through lectures and laboratory experiments, where instructors and students are tied to a textbook for authoritative content information. In recent years, Peer-Led Team Learning (PLTL), Process Oriented Guided Inquiry Learning (POGIL), and project-based learning have been increasingly used by chemistry educators (multiple books and research articles can be found regarding effective use of these pedagogies). Green chemistry as a field is young compared to the more traditional subdiscipline areas of chemistry. As it complements each of these areas, a green chemistry course does not necessarily need to be constrained by prescribed textbook content. Indeed, many educators have been able to incorporate green chemistry principles within traditional subdiscipline courses; resources for integrating green chemistry in this manner can be found in Section 1.5. Freedom to explore a variety of resources and pedagogies makes teaching and learning in green chemistry a somewhat unique sandbox for chemists. Research on teaching and learning indicates that providing an inclusive classroom where students feel like they belong has a great impact on their learning *(10)*. Green chemistry requires scientists to appreciate how chemistry is practiced, and inclusive pedagogy requires instructors to be aware of how they practice their teaching. In addition to inclusion, HIPs and active learning strategies have been shown to engage students in deep and thoughtful ways *(11,12)*.

Starting in 2005, the Association of American Colleges and Universities (AAC&U) launched a 10-year initiative known as LEAP (Liberal Education and America's Promise) that aimed to understand and meet the needs of today's students *(13)*. LEAP has three main initiatives: (1) national public advocacy for education; (2) the campus action network; and (3) research activities. Each of these aim to build a framework for educational changes that promote students' ability to acquire broad knowledge and intellectual/practical skills that allow them to contribute to solving real-world problems. The LEAP initiative was built collaboratively, providing direction about how excellence can be made inclusive *(14)* and leading to the establishment of essential learning outcomes and HIPs *(11)*. In this chapter, pedagogical research from AAC&U, examples of inclusive classroom practices, and discussion of green chemistry from the ACS are all considered and then applied to green chemistry course design.

1.2.1 Essential learning outcomes

Through collaboration with hundreds of colleges and universities, as well as discussion with the business community and input from education-accrediting agencies, a multiyear conversation led by AAC&U resulted in a list of essential learning outcomes that prepare students for success in the 21st century *(15)*. These outcomes transcend any one specific discipline and represent the broad level of education required for our students to tackle new and difficult problems by encouraging high-level thinking, creativity, versatility, and social responsibility. Green chemistry is often thought of as the ethical imperative for chemists *(16)*: it is interdisciplinary and requires adaptation and creativity for solving new (and old) problems. By its nature, a green chemistry course would encompass the majority of these essential learning outcomes. Fig. 1.1 lists these outcomes and ways they can be exemplified as presented by AAC&U. Additionally, examples of applications of each learning outcome in a green chemistry course are provided.

1.2.2 Call for sustainability in education

Instructors immersed in green chemistry education understand that the act of examining and considering our practices through a "green lens" requires us to appreciate principles of sustainability and social responsibility. Sustainability has been described as the "ultimate liberal art (and science)" because it requires asking big questions that necessitate knowledge from a multitude of disciplines ranging from sociology to science and philosophy *(17)*. Indeed, the Association for the Advancement of Sustainability in Higher Education wrote a call to action to create curricula focused on sustainability *(18)*. In addition, hundreds of institutions signed the American College and University Presidents' Climate Commitment, which calls for reforming curricula with an emphasis on sustainability *(19)*. The essential learning outcomes (Fig. 1.1) naturally lend themselves to teaching with sustainability in mind, and integrating green chemistry in the undergraduate curriculum is one way that universities can address this call for reform.

1.2.3 High-impact educational practices

In 2008, George D. Kuh (Chancellor's Professor and Director for Indiana University's Center for Postsecondary Research) prepared a report that defined pedagogical HIPs as teaching methodologies that research indicates have a significant influence on student success *(11)*. These include *first-year seminars and experiences, common intellectual experiences, learning communities, writing-intensive courses, collaborative assignments and projects, undergraduate research, diversity/global learning, service learning/community-based learning, internships, and capstone courses and projects*. While some of these HIPs may not appear to lend themselves toward a green chemistry class, a creative instructor may be able to incorporate several into their course. By gathering data from the National Survey of Student Engagement (NSSE), Kuh established the effectiveness of these HIPs, but also discovered that

Essential Learning Outcomes*	Exemplified by*	Applications in Green Chemistry
Knowledge of human cultures and the physical and natural world	Focusing on engagement with enduring big questions in STEM, humanities and arts	Sustainability ethics Studying chemical life-cycle assessments Multidisciplinary nature of green chemistry
Intellectual and practical skills	Practicing critical thinking, communication, information literacy and teamwork	Group projects Professional presentations Online blogs and video presentations Real-world green chemistry applications Research and community-based projects
Personal and social responsibility	Civic engagement anchored by community involvement in real-world challenges	Community partnerships and collaborations Education outreach Applying green chemistry solutions to local needs Sustainability ethics
Integrative and applied learning	Synthesis of knowledge and demonstration of it through applications to new settings	Creation and dissemination of green chemistry education materials Novel green chemistry research projects

FIGURE 1.1

Essential learning outcomes and potential applications in the green chemistry curriculum.

Adapted from the AAC&U Essential Learning Outcomes National Leadership Council. College Learning for the New Global Century; Association of American Colleges and Universities: Washington, DC, 2007. https://www.aacu.org/sites/default/files/files/LEAP/GlobalCentury_final.pdf.

there was inequity in accessibility of these HIPs for all students. In addition, data showed that students who begin college at lower achievement levels actually benefit more from HIPs than their peers *(11)*. This makes the case for using such HIPs even stronger because they may be most effective for marginalized students. By making HIPs available to all students, we can reach more of them in a meaningful way, which increases inclusivity.

Kuh explains that the HIPs are unusually effective because they "demand that students invest more time and effort into accomplishing a particular activity, result in building relationships between students and faculty mentors, expose students to diversity, provide frequent feedback to students, and require students to see how learning works in various settings" *(11)*. Green chemistry courses that have been described in the literature will be examined in the light of these HIPs, as well as their incorporation of inclusive pedagogy strategies. Whether or not the use of HIPs and inclusive strategies was intentional, a noticeable trend is that a formalized green or sustainable chemistry class lends itself to these impactful teaching approaches.

1.2.4 Call for inclusive excellence

As the undergraduate population shifts to include a higher percentage of first-generation students and those from underrepresented groups *(20)*, there are efforts to examine the traditions entrenched in academia that may not effectively engage this population. Many organizations, including the National Science Foundation, the Howard Hughes Medical Institute (HHMI), AAC&U, and foundations such as Mellon and Lumina are providing research funding and support for institutions to implement inclusive practices. In 2017 and 2018, HHMI awarded 57 Inclusive Excellence (HHMI IE) grants to effect institutional change and increase the capacity for inclusive education beginning in STEM fields *(21)*. These awards call for educators and researchers to fundamentally alter their institutional environment so that all students not only succeed but thrive in their science education. Radford University, where the primary author teaches green chemistry, was one of the 2017 HHMI IE award recipients and the author is involved with inclusive pedagogy faculty development and grant implementation *(22)*.

Based on the HHMI IE grant, design of the Radford University green chemistry course was framed within the realm of "IE." This can be defined in a variety of ways, but the spirit of each definition is to use effective pedagogy to provide every student with the opportunity to excel in science. The notion is that many groups have been marginalized in traditional academic environments, and that this has led to the lack of diversity that is essential for creative and excellent science. By including all students and harnessing their unique abilities, science education will become more effective and equitable.

1.2.5 Inclusive and student-centered teaching practices

What exactly does an inclusive classroom look like? What factors should instructors consider when designing their coursework to ensure that they encompass and build upon the assets of each individual student? While there is not a single approach that will work in the varied classroom situations that exist in higher education, the framework within which inclusive practices reside is common. Salazar et al. suggest that there are five dimensions of IE: intrapersonal awareness, interpersonal awareness, curriculum transformation, inclusive pedagogy, and an inclusive learning environment *(23)*.

Intrapersonal awareness requires faculty to reflect upon their own worldview and recognize how implicit biases may affect their teaching. In comparison, interpersonal awareness encourages the celebration of classroom diversity and encouragement of shared perspectives during class discussions. In terms of intrapersonal awareness, faculty should be prompted to reflect on how their own privilege may influence how they set up their course and interact with students. This may include thinking about the support systems they may have received as a continuing generation college student, or the expectations they have for students' time and availability outside of class. It could also encompass being made aware of microaggressions and reflecting on how participation in the ivory tower of academia has shaped their worldview. To promote interpersonal awareness, faculty can be intentional about providing structured time for students to examine their strengths and weaknesses and to share this with their classmates. For example, the use of strength-based education to create asset maps is a way for students to visualize their own strengths and see how their skills may compliment those of their peers *(24)*. In teaching green chemistry, many instructors use discussion and teamwork, so the ability of the faculty member to encourage students to identify strengths they bring to a team and facilitate open and constructive discussion is critical.

In addition to intra- and interpersonal awareness, Salazar explains that inclusive pedagogy requires curriculum transformation by using a "multicultural lens" to examine curricula *(23)*. One could argue that incorporation of green chemistry into the core chemistry curriculum is transformational in a way that reflects inclusion. Green chemistry has required educators to consider the practice of chemists and industrial processes that have historically prioritized profit over the environment. This naturally leads to social justice discussions and links back to several of the essential learning outcomes listed in Fig. 1.1, such as knowledge of the natural world and personal/social responsibility.

The last two dimensions that Salazar outlines for IE revolve around pedagogy and learning environment. Inclusive pedagogies are teaching strategies that engage the entire student in the learning process. Examples of these include group discussions and debates, student-led discussions, experiential learning, collaborative assignments, and giving students choice and voice in their work. An inclusive learning environment is characterized by a caring instructor who establishes a safe and professional atmosphere in their classroom, where students have a sense of belonging. Focusing on student development and respecting their voice while expecting active participation in the learning process is critical. These inclusive pedagogies can be used in the green chemistry classroom by recognizing students' unique assets, facilitating open group discussions, fostering student choice regarding assignments, using team-based group activities, and providing experiential learning through research or service-learning projects. Many resources are available to learn more about inclusive teaching strategies and many universities have centers for

Resource Title	Created by	Description	Link or Citation
Inclusive Teaching Toolkit	Monash University Library, Australia	Guidelines and tips about inclusive teaching	https://www.monash.edu/library/inclusive-teaching
Theory Into Practice Strategies: Small Groups	Flinders University, Australia	An inclusive approach to facilitating small group work. Includes tips and questionnaire for faculty self-reflection	https://diversity.humboldt.edu/sites/default/files/using_small_groups_to_promote_inclusive_learning_-_flinders_university_australia.pdf
Structure Matters: Twenty-One Teaching Strategies to Promote Student Engagement and Cultivate Classroom Equity	Kimberly Tanner, San Francisco State University	Checklist and description of equitable teaching strategies that faculty can adopt in their classrooms	CBE—Life Sciences Education Vol. 12, 322–331, Fall 2013
Classroom Climate: Creating a Supportive Classroom Environment	Carnegie Mellon University, Eberly Center, Teaching Excellence & Educational Innovation	Provides rationale for considering classroom climate. Links for resources about how to adopt particular aspects of inclusive student engagement	https://www.cmu.edu/teaching/designteach/teach/classroomclimate
Learner-Centered Teaching: Putting the Research on Learning into Practice	Terry Doyle, Learner-Centered Teaching	Cognition and learning science presented in the context of moving the teacher to the role of facilitator and developing students as active learners	Stylus Publishing, LLC. ISBN-978-1-5792-2742-5
The Diverse and Inclusive Classroom	The Learning & Teaching Office, Ryerson University, Canada	Tips on reaching diverse students, encouraging discourse, awareness of student needs and resources	https://www.ryerson.ca/content/dam/lt/programs/workshops/Diverse_Inclusive_Classroom.pdf
Breakthrough Strategies: Classroom-Based Practices to Support New Majority College Students	Kathleen A. Ross, Heritage University	Teaching strategies for engagement, promoting sense of belonging, engendering confidence and building student professional identity	Harvard Education Press. ISBN: 978-1-31250-997-6

FIGURE 1.2

Resources for inclusive pedagogical practices with links and descriptions.

teaching and learning that have support in this area. Fig. 1.2 provides a few inclusive pedagogy resources to show the breadth of ways that faculty can be inclusive in their classroom.

1.2.6 The American Chemical Society's role in green chemistry and pedagogy

The required curriculum for a United States chemistry degree is greatly influenced by the ACS's CPT. As of October 1, 2018, there are 693 bachelor's degree granting institutions that have a curriculum approved by the ACS. Green chemistry is mentioned twice in the current iteration of CPT's *ACS Guidelines and Evaluation Procedures for Bachelor's Degree Programs (9)*. In their description of in-depth coursework, CPT urges programs to incorporate green/sustainable chemistry as a modern topic. Green chemistry is also mentioned in the ethics section that describes the necessity of exposing students to chemistry's role in current global and societal issues. The ACS Committee on Environmental Improvement contributed the *Green Chemistry in the Curriculum* supplement to the ACS CPT Guidelines in March 2018 *(25)*. This outlines green chemistry conceptual and practical topics that can be addressed within coursework. The numerous resources available from the ACS illuminate the importance chemists place on green chemistry in education. Therefore, while the ACS has not explicitly *required* incorporation of green chemistry into undergraduate curricula through their CPT guidelines, the ACS is encouraging implementation of green chemistry in teaching through providing resources, networking, and grant opportunities. In terms of pedagogy, the ACS promotes the continual examination and enhancement of teaching practices: "Programs should teach their courses in a challenging, engaging, and inclusive manner that accommodates a variety of learning styles" *(9)*. Separately from the CPT guidelines, the ACS also has a diversity and inclusion statement *(26)* and symposia such as "Inclusive Excellence in Academic Leadership" at national meetings indicating that progress toward inclusion is being made.

1.2.7 Resistance to change

Inclusive pedagogies and incorporation of green chemistry principles have met with resistant arguments from academics. Inclusive teaching focuses on understanding and meeting the needs of the changing undergraduate population, instead of trying to "fix" the student. However, many academics still use deficit language when describing their students. In parallel, some chemists have resisted the call to teach green chemistry principles, proclaiming that greening the curriculum will leave students unprepared to work in industrial environments. In the field of biology, Brownell and Tanner address some of the barriers to pedagogical change, and the tensions that innovative teaching has with professional identity *(27)*. These parallels are drawn to show that it will take transformative change for educators to teach with inclusion, redesign curricula, and wholeheartedly embrace the ideals of green chemistry. As instructors adopt inclusive teaching practices and document increased success of their first-generation and nonmajority students, this evidence can be leveraged to engage those faculty who are resistant to change.

1.3 Green chemistry courses incorporating impactful and inclusive pedagogies

Green chemistry courses for major and nonmajor students differ from the typical prescribed subdiscipline courses in chemistry and often allow for more creativity in content delivery. In this way, green courses may showcase how teaching can be inclusive and active, as opposed to traditionally lecture-based, and be used as vehicles for real curricular impact. As described in Section 1.2, the ideals of green chemistry align well with the incorporation of essential learning outcomes and more active and inclusive teaching. The description and design of several green chemistry courses have been published in the chemistry education literature during the last 15 years. A few of these courses are highlighted here in the context of how each instructor incorporated inclusive, active learning, and/or high-impact pedagogical practices. Although inclusive pedagogy and HIPs are not synonymous, using certain approaches such as service-learning and collaborative projects are likely to increase students' sense of belonging, which is a tenant of inclusive pedagogy. Active learning gives students the responsibility of acquiring the knowledge they need and encourages them to take ownership of their own learning, which necessarily includes them in the classroom. Essential learning outcomes, inclusion, HIPs, and sustainability have been successfully weaved together to create meaningful green chemistry courses in higher education.

1.3.1 Toward the Greening of Our Minds: An upper-level green chemistry course for science majors and minors (7)

One of the earliest examples of a green chemistry course published in the education literature was from Anne Marteel-Parrish, where she described teaching a special topics course specifically using the "learning now" method. This style uses student engagement and active learning that requires students to do intellectual work to obtain and interact with new knowledge. In the first course section, the instructor established the expectation for participation by requiring intellectual contributions throughout class that prepared students in the basics of green chemistry. Following this, the second part of the course gave students a choice in researching a real-life example of green chemistry and required them to organize a class discussion on their topic. This required students to utilize background theory that they learned and to apply it to a new topic, while placing them at the helm of the class. In the third section, students researched a Presidential Green Chemistry Award of their selection, and then wrote a paper and prepared an oral presentation and discussion about the award. These methods encouraged students to be integral in choosing the direction of the course and empowered them to be actively engaged. The final part of the class required students to collaborate in pairs to create a unique green chemistry solution to a synthesis or process, to write an abstract based on their solution, and to prepare an oral presentation on their work. Overall, students indicated via written comments that the active learning portions were much more effective than the lecture components of the class. They additionally noted that they enjoyed the student-directed aspects of the course.

This course successfully incorporated the essential learning outcome of developing students' intellectual and practical skills. Through active participation in discussions, professional presentations, and group research projects, students in this course exemplified critical thinking, information literacy, and teamwork in problem-solving. By providing student choice and empowering student voice, the instructor worked to create an inclusive learning environment. This early instance of a green chemistry course has provided an excellent example to inspire other green chemistry educators.

1.3.2 Green Chemistry and Sustainability: An honors science course *(3)*

An honors course for both science and nonscience majors, taught by Erin Gross at Creighton University, was designed with a multidisciplinary approach. The course was organized around the three central themes of energy, pollution and waste prevention, and safety. Gross intentionally scaffolded student learning through teaching the basics of green chemistry to ultimately requiring original research proposals. Additionally, scientific literacy was developed for students by firstly performing literature searches, then creating a literature review, and finally writing a scientific proposal. This class incorporated the HIP of intensive writing as students were required to write original research proposals modeled after the Environmental Protection Agency's (EPA's) P3 program. They also were required to keep "green journals" that encouraged them to become more aware of their environmental impact by recording and reflecting on their readings. The instructor commented that the journals allowed her to get to know the students, especially the quieter ones, indicating that she cares about them (an essential element in creating an inclusive environment) *(23)*. Students contributed to and learned from their community in the form of field trips and participation in the Conversations Conference on Nebraska Environment and Sustainability, which aligns well with the essential learning outcome of personal and social responsibility. Inclusive pedagogy literature discusses the importance of helping students develop a professional identity *(28)*, which is certainly enhanced by activities such as participating in the state conference and writing an original research proposal.

1.3.3 Green Chemical Concepts: A course for nonscience majors *(6)*

This nonmajors lecture and laboratory course at Susquehanna University specifically taught sustainability from a chemistry perspective to students that would potentially be making future policy or business decisions. The instructor (Renuka Manchanayakage) used interactive lectures where participation was expected so that students could share their discipline-specific perspective on green chemistry concepts. Students undertook laboratory work to learn modern chemistry techniques and engaged in an atom economy workshop to help them become aware of pollution and waste prevention. Additionally, teams of students learned how to collaborate in creating a life-cycle assessment to compare two products or processes in terms of their green qualities, and collectively presented their research.

By serving nonscience majors, Manchanayakage was able to expose these students to knowledge of the physical and natural worlds and encouraged them to learn practical skills around communication, information literacy, and teamwork, all of which are AAC&U essential learning outcomes. The collaborative life-cycle assessment project is an example of an HIP utilized in this course. Additionally, the instructor encouraged students to discuss their own unique viewpoint on each of the green chemistry topics that were introduced. Including student voices and facilitating them to speak from their experiences is an example of how to create an atmosphere of inclusion. The positive student evaluations of this course led to the development of an offering for major students, as discussed in the following section.

1.3.4 Green Chemistry: A course for science majors (6)

An upper-level elective green chemistry course, also designed by Manchanayakage, was taught in a workshop style for students to move fluidly between lecture and laboratory activities. It was designed to integrate many science disciplines (chemistry, biochemistry, ecology, and earth and environmental science) as a model for the interdisciplinary, collaborative work that drives scientific discovery. Students were empowered to share their expertise through dialogue on green chemistry topics. Critical thinking was promoted via the evaluation of alternative synthetic methods by applying green chemistry metrics. The HIPs and elements of inclusive pedagogy used in this course were evidenced by the collaborative projects and teamwork components. Students developed and orally presented an original proposal for designing a green product or process that was relevant to their major, which exemplifies the essential learning outcome of integrating and applying their learning.

1.3.5 Green Goggles: A nonmajors green chemistry course (8)

This nonmajors course, designed and taught by Sarah Prescott at the University of New Hampshire (UNH), is part of the UNH Discovery curriculum where students develop unique strategies to address unanswered questions, present their research, and reflect on their learning. The course taught fundamental chemistry concepts through the lens of green chemistry principles and practice. It met for 3-h blocks and therefore the instructor designed the course to engage students with active learning strategies. Students were required to prepare for class by reading Stanley Manahan's free online textbook (29) and then participated in active learning and authentic discussion sparked by their reflections on what they discovered in their independent reading between the weekly class meetings. "Just-in-time" teaching was utilized to teach scientific concepts as they emerged from discussions of green chemistry. In-class assignments included collaborative problem-solving that helped students to learn information that would be required for successful completion of each activity. Graded green chemistry blogs provided a format for particularly introverted students to actively participate in class discussions. Students also created a wiki site that demonstrated connections between general chemistry and green chemistry concepts based on a particular topic. This required them to create a product (e.g., a video or podcast) that could be used at the conclusion of the course. The instructor scaffolded this project-based learning over the entire semester and required students to write a formal paper and present orally at the UNH Science Symposium. These varied assessment types allowed students to demonstrate their skills in different ways, and the utilization of "just-in-time" teaching and collaborative projects indicate that the instructor focused on reaching each individual in the course. Upon reflection on the format of having two heavily weighted examinations in this course, the instructor commented that future sections would have more frequent and slightly lower-stakes assessments in the form of quizzes.

Course evaluations using the Student Assessment of Their Learning Gains (SENCER-SALG) showed that students felt they made the most improvement when participating in discussion, group work, and hands-on classroom activities. Many students gave feedback that indicated they appreciated the variety of ways that they could learn in this course. The instructor provided a "comfortable" environment for students to participate in discussion, which helped increase their interest in the chemistry content. Students reported that this course was an accessible way to learn science, and they made correlations with science, ethics, and society. The varied teaching strategies in this course and the care taken by the instructor to set up a supportive learning environment align with AAC&U's essential learning outcomes and inclusive teaching practices. This course design is an excellent example of effective pedagogy applied in green chemistry.

1.3.6 Green Chemistry and Sustainability: An online upper-level green chemistry course (4)

A new online green chemistry course, designed by graduate students at the University of Cincinnati, targeted students who had a background in general and organic chemistry. Over half of the enrolled students were chemical engineers and the other half were students in biology, chemistry, or environmental studies majors. The course content was delivered in 15 online modules where the principles and applications of green chemistry were explored. A module typically included a video to introduce a topic followed by self-guided assignments to apply and demonstrate new knowledge. Some of the assignment types included posting to discussion boards, written summaries or reflections, guided reading comprehension quizzes, group work, and a final written paper. Regarding inclusive pedagogy, the course instructors provided almost immediate feedback, material was "chunked" and distributed online on a weekly basis to avoid overwhelming the students, and a variety of assessment types were offered. The instructors were available via online office hours to assist students who had questions about the material. They encouraged virtual meetings to help expose students to this common communication method in preparation for their future careers.

1.3.7 Collection of additional course syllabi showcasing green chemistry

In addition to the outstanding examples of green chemistry courses described here, a collection of syllabi from undergraduate courses that have significantly incorporated green chemistry are indicated in Fig. 1.3. Citations are listed so that instructors who are seeking ideas for their own class material, learning objectives, or curricular structure can explore a variety of resources. Green chemistry courses and syllabi were found by searching articles in the *Journal of Chemical Education* and by searching for courses offered at colleges and universities that have joined the Beyond Benign Green Chemistry Commitment *(30)*. All syllabi are from classes that have been offered during the last 8 years.

Institution	Author	Course Type	DOI or URL for Syllabus
KTH Royal Institute of Technology	Brian J.J. Timmer & Fredrik Schaufelberger	Organic Laboratory	DOI: 10.1021/acs.jchemed.7b00720
Oral Roberts University	Lois Ablin	Organic Laboratory	DOI: 10.1021/acs.jchemed.7b00570
University of Toledo	Mark R. Mason	Green Chemistry	http://www.utoledo.edu/nsm/chemistry/pdfs/BBBCHEM4200%20Syllabus.pdf
North Carolina State University	Lucian Lucia	Green Chemistry	https://faculty.cnr.ncsu.edu/lucianlucia/wp-content/uploads/sites/11/2016/03/GC-Course-Syllabus-14Sept2015.pdf
Washington College	Anne E. Marteel-Parrish	Green Chemistry	DOI: 10.1021/ed400393b
Michigan Green Chemistry Clearinghouse	Multiple authors	Green Chemistry, Green Engineering, Toxicology	http://www.migreenchemistry.org/education/classroom-resources/syllabi-and-lecture-materials

FIGURE 1.3

Examples of syllabi from courses focused on green chemistry.

1.4 A case study in green chemistry course design

1.4.1 Special Topics: Green Chemistry, an upper-level elective course for science majors

Considering inclusive pedagogy, HIPs, and principles of "backwards course design," a new upper-level green chemistry class was designed at Radford University. This course was taught by Sarah Kennedy and taken by Rachel Chapman (the coauthors of this chapter), and modified from the Kennedy course at Westminster College *(5)*.

The course design was presented at the 2018 ACS Biennial Conference on Chemical Education and subsequently selected for inclusion in this book due to the focus on green chemistry and pedagogy. In the true spirit of active learning and real-world experience, Rachel was invited to coauthor this chapter to provide a student perspective. Here, we outline the course as it is taught at Radford University and how it connects with the pedagogical principles outlined previously. (Although the course materials are not provided here, the corresponding author is willing to share instructional resources via email.)

1.4.2 Course overview

"Green Chemistry" at Radford University is an upper-level course designed as an elective for chemistry major and minor students. Radford University is a midsize public comprehensive state university, where the Department of Chemistry is certified by the ACS. The prerequisite for this course is Organic Chemistry I with a corequisite of Organic Chemistry II. The expected majority enrollment is from junior and senior chemistry majors, or biology majors with a chemistry minor and interest in environmental chemistry. The course was taught for the first time during the Spring 2018 semester with an enrollment of eight students, and met for 75 min twice per week for 14 weeks with an additional final examination (2 h in length).

Since many chemistry majors have only a few chances to take upper-level elective courses, this course was designed to focus on the students' overall professional development by using HIPs, inclusive pedagogy, and focusing on enduring essential learning outcomes. One of the instructor's goals was to be transparent about the reasons for the course design and the purpose for each assignment and project. Students were often given choice in their project design, assessment methods, and "teammates." The instructor emphasized that this course would be driven by content, but that there would be a variety of assignments that would focus on developing themselves as professionals, including teamwork, writing, oral presentations, and independent research.

Instead of a traditional teacher-student model classroom, the instructor strove to create a student-scholar atmosphere. Many of the ideas embedded in the essential learning outcomes, HIPs, and inclusive pedagogy focus on teaching holistically and applying knowledge to real-world challenges, which are important for students to become independent thinkers and professionally competitive. Therefore, the instructional approach was focused on the students' professional development while also being informed by many evidence-based teaching practices. With the course enrollment of eight undergraduates, student "buy-in" was essential to elicit the most productive outcomes for the semester. As such, first day activities were especially important to set the stage for an interactive class atmosphere.

1.4.3 **Course objectives and timeline**

The course overview was provided in the syllabus and discussed at the outset of the first class:

Green chemistry is the study of the principles, concepts, and applications of sustainable chemistry. Historical context will provide an understanding of the reason why green chemistry is so important. The 12 Principles of Green Chemistry and their application in chemical industry will be explored through readings, discussions, case-studies and projects. Particular attention will be given to industrial processes, green metrics, catalysis, solvents, and renewable resources.

Using "backwards design," the desired outcomes were considered when creating the course learning objectives. These were written in a format for students to be able to appreciate what they should be able to do at the end of the semester and acted as a guide for the course.

At the end of this course, students should be able to…

- *describe the importance of green chemistry based on the historical context of the Industrial Revolution and chemical accidents.*
- *demonstrate an understanding of the 12 Principles of Green Chemistry.*
- *utilize metrics to evaluate processes in terms of their greenness.*
- *critically evaluate current research in green chemistry and demonstrate understanding of award-winning green chemistry projects.*
- *articulate the importance of all chemists incorporating green principles in their projects.*
- *contribute to educating others about green chemistry.*

With these learning objectives in mind, the course literature was selected, the main projects were designed, and a timeline of the course was created. However, in the spirit of the student-scholar model and providing students with voice and choice, several of the project details and reading assignments were left ambiguous to allow for student input throughout the semester. Additionally, students were often provided with initial readings and then required to utilize other resources to deepen their understanding of the content and application of each topic. The class met on Tuesdays and Thursdays with a pattern of reading and discussions on Tuesdays to introduce a topic, then diving deeper through projects, case studies, or primary literature exploration on Thursdays.

1.4.4 **First day activities**

The first day of the course was crucial to set the stage for team-building and student participation. Students first worked in pairs to learn about each other's career goals and their interest in green chemistry. To engage them in thinking about science as a discipline that requires community, they also reflected about two people who inspired them during their academic career. Each student then introduced their partner to the class in an informal circle, which would be the room configuration for the weekly discussions.

After the introductions, there was a short discussion with slide prompts about the course learning objectives and specifically how this upper-level class would be different from other science classes. A discussion of the top traits that employers desire was discussed so students could see the value in teamwork, communication styles, critical thinking, problem-solving, and information literacy that they were going to be engaged in during the semester. This immediately and intentionally led to a discussion about how the class might help them gain valuable employable skills in the context of learning and applying green chemistry. This helped to promote their sense of belonging and professional identity as scientists *(28)*.

Since the group would work in small teams throughout the semester on various projects, the marshmallow challenge *(31)* was used as a way for them to work together toward a common goal. For 15 min, two teams of four students had 50 mini-marshmallows and 100 toothpicks with which to construct a tower that was to stand eight inches tall and support the weight of a methane molecule model for 10 s. The true aim of the challenge was not for them to create the tower successfully, but to actively work together on a project with low stakes. The task required them to evaluate the problem at hand, to plan and build a prototype, and to develop communication within the team. After completion of the task, the students were asked reflective questions to examine how their team worked together and what role they individually played. They discussed how leadership emerged on their team, how they kept track of the team member's ideas, how decisions were made, how time factored into their process, and what they might be able to take from this challenge and apply to how they work within a group. The challenge provided a common learning experience that could be referenced in future class meetings.

The first day closed with a discussion of how the content of this course and the pedagogy used were designed to help them build on their chemistry knowledge and apply it to new problems in a collaborative manner. The course required them to be active participants, who were expected to come prepared and participate in each class session and project. Quite remarkably, the class had 100% attendance for the entire semester, which is a testament to how committed they were to learning green chemistry and receptive they were to the style of instruction.

1.4.5 Including student voice and student choice

While the skeletal outline and timeline for exploring green chemistry topics was provided in the course schedule, there was intentional vagueness in the assignments so that students were able to have input on the types of projects they wanted to explore in the classroom. During the first few weeks, historical chemical disasters were explored to underscore the imperative of green chemistry. In this section, students researched the environmental damage done by chemical accidents and presented their work in poster format. They were able to select the topic themselves and were guided through a literature search with an instructional librarian. In this era of "fake news," the library session aimed to explore a variety of online news and literature sources to teach students how to discern the quality and reliability of available information.

After this activity, students undertook a short reflective writing assignment. Firstly, they were asked to reflect on their class participation by looking at the rubric for the course and grading themselves on their contributions. A second question prompted them to think about the balance between the negative effects of chemical disasters they had just researched and the positive ways that chemistry influences our everyday lives. Input was requested regarding how to showcase the positive aspects of chemistry and what the next assignment might look like to strike a balance with their chemical disaster project. Finally, the students were asked what kind of project-based learning they might like to see in the classroom and how they thought it could be implemented. From these questions, the class discussed what the next assignment should entail and how it would be constructed. Students were polled several more times throughout the semester to evaluate themselves on their class contributions and to assist in guiding the direction of future projects and assessments.

1.4.6 Projects and assessments

Inclusive teaching resources discuss the importance of having varied types of assessments so that students have multiple formats to demonstrate their knowledge *(32)*. In this course, students were expected to participate in class discussions, but they also had multiple other ways to showcase knowledge including creation of posters, brochures, oral presentations, concept maps, and a community-based or laboratory-based final project. Activities throughout the semester focused as much as possible on the what, why, and how of green chemistry. For example, one project provided freedom to design educational materials that could teach the basics of green chemistry to a wide audience. This gave students some latitude, encouraged them to research real-world applications of green chemistry, and required them to be able to communicate their findings to nonscientists. Several students used this as an opportunity to research current applications of green chemistry in industries that were interesting to them, including athletics and polymers. The final project required collaboration with teammates to either teach a laboratory with green chemistry principles or undertake some novel research. Ideas for this evaluation were not dictated by the instructor but were reached upon consensus after discussing various possible ways that new green chemistry knowledge might be applied. Student groups wrote their final project proposals along with their suggested evaluation (grading) plan and met with the instructor to approve the project idea, assessment plan, and timeline. Three groups were formed, and the projects included (1) creating and teaching a high-school green chemistry laboratory; (2) creating and teaching a general chemistry undergraduate laboratory; and (3) designing an experiment to monitor the uptake of lead by mushrooms.

The final course grade was composed of project work (40%) and class participation, preparedness, and in-class activities (60%). A rubric was used for participation where a judgment was made based on the quality of contributions made by each student and how they enhanced or detracted from the class discussion. Each student suggested a score for their participation several times throughout the course and defended their score with evidence, which encouraged metacognitive reflection on their contributions. Conversations regarding their self-evaluations led to discussions about the value of giving an opinion in class versus providing additional evidence for arguments based on research.

1.4.7 Student evaluations and perspectives on course design

Formal evaluations indicated that students felt they learned a great deal in the course and that they were required to think deeply about the content. Several comments indicated that students appreciated the focus on their professional development and allowing them choice in their assignments. The most-criticized part of the course was the apparent disorganization of how assignments would be graded. In retrospect, allowing students the choice on their evaluations is valuable for ownership of their learning, but once input is received, the evaluative measures should be formalized and agreed upon so that there is no confusion about expectations.

In addition to the standard evaluation, the students were asked a series of questions based on the learning objectives and the course design. When prompted to provide any general comments or suggestions in these areas, three out of eight students referenced that they liked that they could choose their projects:

- "I liked the open course design with minimal structure and choosing project topics."
- "I enjoyed how the course was set up, and that we had a say on what we wanted to learn and how we were graded."
- "Letting us choose our own topics was great!"

For this same question, three students had no additional comments, but two also reflected that they liked the nontraditional aspect of the course:

- "Great class, lots of readings and projects."
- "Loved debating with other students in class."

However, one student did not appreciate this nontraditional approach to a chemistry course and provided comments indicating that they would like more structure, direction on assignments, and that the design was "a little loose for a 400 [level] course." The course design was considered effective in meeting the course objectives and promoting professional skills (Fig. 1.4).

Overall, the first green chemistry course at Radford University met the course objectives while being designed with inclusive and impactful teaching strategies. In future iterations the instructor plans to use many more of the available online resources for green chemistry (outlined in Section 1.5) and empower the students to more formally share the knowledge that they gained in the course through presentations at internal or external venues.

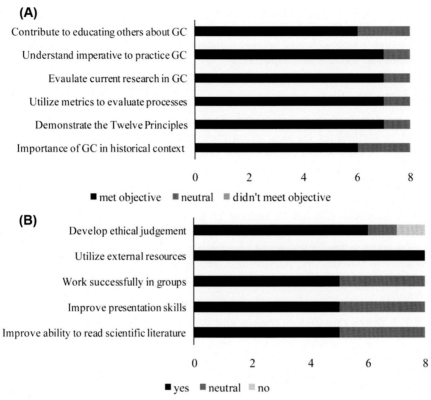

FIGURE 1.4

Student evaluations regarding course objectives and design for "Green Chemistry." (A) Course objectives: student responses to "Did we meet the objectives of the course?" with response choices of (1) met objective; (2) neutral; and (3) did not meet objective. Each question reminded them of an objective from the syllabus. Following each rating, students were asked to provide evidence for their selection. (B) Course design: student responses to "Was the course design effective?" with response choices of (1) yes; (2) neutral; and (3) no. Each question asked about a specific skill related to their professional development.

1.5 Collection and review of resources for teaching green chemistry

Since 1998 when Anastas and Warner first published on the theory and practice of green chemistry *(1)*, numerous books, journal articles, and case studies have been written on the topic. Organizations such as the ACS, the EPA, and Beyond Benign have incentivized scientists to engage in green chemistry and to develop applicable curricular materials. In this section, we highlight some textbooks, supplementary educational resources, and web resources that may be useful for instructors seeking to integrate green chemistry into their existing courses or to develop their own green chemistry course. Materials discussed here may be used by educators from introductory through advanced levels, ranging from general chemistry to organic, engineering, industrial and research applications.

1.5.1 Textbooks

A variety of textbooks have been authored by chemists, engineers, and educators that provide a scaffold for teaching green chemistry (Fig. 1.5). Some texts, such as *Chemistry in Context* and *Chemistry for Changing Times* are specifically aimed at students who are at the introductory level. A book such as Anastas and Warner's *Green Chemistry: Theory and Practice* can be read by nascent chemists as an inspiration to practice green chemistry. Albert Matlack and Mike Lancaster offer examples of textbooks that can be utilized in a green chemistry class with organic chemistry as a prerequisite. Several other textbooks are available that have specific foci on green technologies, green engineering, or other aspects of sustainability. Examples of introductory, intermediate, and advanced textbooks are provided to give an overview of the variety and level of textbooks available to teach green chemistry.

1.5.1.1 Green Chemistry: Theory and Practice by Paul Anastas and John Warner (1)

This book is an introductory text to the modern environmental sustainability movement and the ethical imperative of practicing green chemistry. It represents where Anastas and Warner first published the 12 Principles of Green Chemistry that have become the guiding principles in the field. The first chapter describes common misconceptions of safe chemical use and disposal. Subsequent chapters describe the 12 Principles and how they can be used as guidelines for methodology and evaluation. This textbook also introduces some chemical concepts such as reaction mechanisms and catalysis that some readers may not have encountered before. It succeeds in setting the stage for the green chemistry imperative and whets the appetite of the reader to explore the principles in a deeper and more meaningful way.

1.5.1.2 Green Chemistry: An Introductory Text by Mike Lancaster (33)

Providing a solid framework for a green chemistry course, this book begins by establishing the foundational concepts of green chemistry and highlights the impacts of the chemical industry. Overall, it encourages an understanding of the inner workings of green chemistry in research, industry, ethics, technology, governmental policy, resources and feedstocks. The book includes detailed explanations of processes used in these different fields and relates them to sustainability and introduces the reader to many industrial aspects, including reactor design, energy input, chemical life cycles, and scale-up. Process monitoring and the importance of implementing safety within each process step is also highlighted. The focus on industry may be the first time that

Title (Edition)	Author or Editor(s); Publisher, Year	Level
*Green Chemistry: Theory and Practice	Paul Anastas and John Warner; Oxford University Press, USA, 1998	Introductory
Chemistry for Changing Times, 13th Edition	John W. Hill, Terry W. McCreary, Doris K. Kolb; Prentice Hall, 2012	Introductory
Chemistry in Context, 8th Edition	American Chemical Society; McGraw-Hill Higher Education, 2015	Introductory
Green Chemistry and the Ten Commandments of Sustainability	Stanley E. Manahan; ChemChar Research, Inc., 2006	Introductory
*Green Chemistry: An Introductory Text, 2nd Edition	Mike Lancaster; Royal Society of Chemistry, 2013	Intermediate
An Introduction to Green Chemistry, 2nd Edition	Albert Matlack; CRC Press, 2010	Intermediate
Green Chemistry Fundamentals and Applications	Suresh Ameta and Rakshit Ameta; Apple Academic Press, 2013	Intermediate
Key Elements of Green Chemistry	Lucian Lucia; North Carolina State University Independent Publisher (free online), 2018	Intermediate
*Green Chemistry: An Inclusive Approach	Bela Török and Timothy Dransfield; Elsevier, 2018	Advanced
Green Organic Chemistry and its Interdisciplinary Applications	Vera M. Kolb; CRC Press, 2016	Advanced
Green Chemistry for Organic Synthesis and Medicinal Chemistry	Wei Zhang, Berkeley Cue; Wiley, 2012	Advanced
Handbook for Green Chemistry and Technology, 1st Edition	James H. Clark and Duncan Macquarrie; Wiley-Blackwell, 2002	Advanced
New Trends in Green Chemistry	V. K. Ahluwalia and M. Kidwai; Springer, 2004	Advanced
Green Chemistry and Engineering: A Pathway to Sustainability	Anne E. Marteel-Parish and Martin A. Abraham; Wiley, 2013	Introductory, engineering
Green Chemistry and Engineering: A Practical Design Approach	Concepción Jiménez-González and David J.C. Constable; Wiley, 2011	Intermediate, engineering

FIGURE 1.5

Selected textbooks available for use in teaching green chemistry organized by academic level. "Introductory" can be used in general chemistry, "intermediate" can be used with an organic chemistry pre/corequisite, and "advanced" level includes more challenging concepts often with industrial examples. *Titles beginning with an asterisk have been summarized in this chapter.

a traditional science major becomes exposed to the concepts of scale-up and engineering processes. Each chapter ends with review questions and supplemental readings that can be used to initiate deeper research and discussion.

1.5.1.3 Green Chemistry: An Inclusive Approach edited by Bela Török and Timothy Dransfield (34)

This book is divided into three main sections, providing a general background in green chemistry and then numerous unique examples of applications. The first two parts follow a textbook layout to introduce the basic principles of green chemistry, environmental chemistry, toxicology, and sustainable synthesis. These chapters include critical thinking questions and recommended supplemental literature. This text was classified as advanced in Fig. 1.5 because of the extensive third part that explores 39 different applications of green chemistry. This section, which accounts for the majority of the book, explores practical green chemistry by outlining topics such as composition of the natural world, toxicology, synthetic design methodologies, green research initiatives, and green engineering. Each of these is grounded in research literature or industrial applications to capture some innovative ways to apply green chemistry principles in various fields.

1.5.2 Journals featuring green chemistry

Several peer-reviewed journals focus on green chemistry or contain specific sections featuring green or sustainable chemistry. The ACS has numerous publications where applications of green chemistry in each subdiscipline may be found. Of interest are the journals *Environmental Science and Technology* and *Journal of Chemical Education*, both of which have many articles on topics that can be brought into the classroom. The Royal Society of Chemistry's *Green Chemistry* and Elsevier's *Current Opinion in Green and Sustainable Chemistry* are also very rich resources for finding current research that can be explored by students.

1.5.3 Instructional activities and case studies in green chemistry

1.5.3.1 Introduction to Green Chemistry: Instructional Activities for Introductory Chemistry edited by Mary Ann Ryan and Michael Tinnesand (35)

Instructional activities for introductory chemistry produced by the collaborative effort of the chemical societies of Germany, the United Kingdom, and the United States are collected in this text. This serves to teach the fundamentals of a general chemistry course through the lens of green chemistry and sustainability. Each chapter is linked to one of the 12 Principles of Green Chemistry and describes procedures for several activities. Instructional notes for each activity, helpful diagrams, question sets with descriptive answers for instructors, reference materials, and discussion points to increase the depth of knowledge on each topic are provided.

1.5.3.2 *Real-World Cases in Green Chemistry edited by Michael Cann and Thomas Umile* (36)

This work is a collection of case studies that have been created based on the Presidential Green Chemistry Challenge Awards *(37)*. Some feature popular introductory topic exercises (e.g., atom economy calculations) and connect real-world applications to green chemistry solutions. Full case studies are provided with an overview, direct problem, and the green chemistry solution. The overview gives background information for the case study topic and how the solution to the problem successfully employed green chemistry. At the end of each case, reflection questions are provided. For the instructor, notes about how to teach each example, the relevant field of chemistry to which the study pertains, and related discussion topics are presented. This collection challenges students to explore external resources to answer questions and connect ideas, which promotes knowledge synthesis. The book recommends that the student should have a background in organic chemistry to understand terminology, reaction mechanisms, and synthesis.

1.5.4 Green chemistry laboratory manuals and methodologies

1.5.4.1 *Greener Approaches to Undergraduate Chemistry Experiments by Mary Kirchhoff and Mary Ann Ryan* (38)

This laboratory manual contains 14 experiments featuring introductory chemistry concepts. Each procedure is formatted with clear sections that list the corresponding green chemistry topics, introductory material, reaction schemes, prelaboratory questions, cumulative postlaboratory questions, and specific safety precautions to follow during procedures. Instructional pointers for experiment setup and cleanup are provided, along with green assessments and green opportunities for reaction improvements.

1.5.4.2 *Green Organic Chemistry: Strategies, Tools, and Laboratory Experiments by Ken Doxsee and Jim Hutchison* (39)

The lessons and experiments presented in this text are from the green organic chemistry curriculum at the University of Oregon that has replaced the more traditional organic laboratory. The intention is for this book to be used as a stand-alone green chemistry laboratory manual or in conjunction with another manual. The first section is dedicated to strategies and tools following the 12 Principles of Green Chemistry that can be implemented into laboratory lessons, such as identification of chemical hazards, using alternative solvents and reagents, reaction efficiency, exploring alternative feedstocks, and employing green chemistry metrics. The second section is comprised of 19 laboratory experiments that each contain prelaboratory preparation, a reaction procedure, postlaboratory questions/exercises, developmental notes, and related chemistry concepts/techniques. The experiments in this book were taught to numerous instructors during week-long workshops at the University of Oregon for many years as part of the NSF-sponsored Center for Workshops in the Chemical Sciences (CWCS) program.

1.5.4.3 Experiments in Green and Sustainable Chemistry edited by Herbert Roesky and Dietmar Kennepohl (40)

This resource is a collection of laboratory experiments from many contributors that highlight chemistry's role in energy and material resource sustainability. It contains procedures for 46 experiments in five areas: catalysis, solvents, high-yield and one-pot synthesis, limiting waste and exposure, and special topics. An introduction for each section includes foundational concepts, importance of the topic, usefulness, history, and how processes have changed and improved over time. Each experiment typically includes the following sections: introduction, apparatus, chemicals, risks, hazard precautions, procedure, waste disposal, points for observation, explanation of phenomenon and references.

1.5.4.4 Methods and Reagents for Green Chemistry: An Introduction by Alvise Perosa and Fulvio Zecchini (41)

This text was developed from the 1998−2003 lectures presented at the Summer School for Green Chemistry in Venice, Italy. Overall, unique procedures using green reagents, alternative reaction conditions, and catalysis are described. The chapters cover subjects such as multicomponents reactions, carbohydrate materials, photoinitiated synthesis, reactions conducted in green reagents, designer solvents, incineration processes, and many types of catalysis including biocatalysts for industrial processes. The subjects are explained in detail with synthesis diagrams, analytical techniques, and figures to describe processes. This book is meant to be used as a laboratory technique guide for chemists and chemical engineers in graduate and postgraduate studies.

1.5.5 Institutions providing online green chemistry educational resources

In addition to the print resources outlined, there are numerous online resources for green chemistry education. These range from educational tools for K-16 to research tools useful for chemists that wish to assess their work through a green lens. Fig. 1.6 lists a variety of current green chemistry online resources with in-text explanations. Due to the wealth of green chemistry tools available from some of the larger institutions, simply a preview of these tools is included here. Resources were identified by searching the ACS's list of universities that signed the Beyond Benign Green Chemistry Commitment (30).

1.5.5.1 American Chemical Society Green Chemistry website

The ACS Green Chemistry website includes countless resources that can be used in education and provides tools of interest to professional chemists. The website is

Organization	Level	Type of Resource	Website
American Chemical Society	All (K-16)	Green chemistry homepage	https://www.acs.org/content/acs/en/greenchemistry.html
Beyond Benign	All (K-16)	Homepage	https://www.beyondbenign.org/k12
United States Environmental Protection Agency	Introductory to advanced	Green chemistry homepage	https://www.epa.gov/greenchemistry
Michigan Green Chemistry Clearing House	Introductory to intermediate	Homepage	https://www.migreenchemistry.org/education/classroom-resources
Center for Green Chemistry & Green Engineering at Yale University	Introductory to advanced	Online learning modules & green chemistry videos	https://greenchemistry.yale.edu/education/undergraduate-graduate/modrnu-modules
University of Scranton	Introductory to advanced	Online modules for greening various types of chemistry courses	http://www.scranton.edu/faculty/cannm/green-chemistry/english/drefusmodules.shtml
The Institute for Green Science	Introductory to advanced	Online green chemistry course, videos, learning modules	http://igs.chem.cmu.edu

FIGURE 1.6

Institutions providing green chemistry web resources. "Introductory level" indicates general chemistry, "intermediate level" indicates prerequisite of organic chemistry, and "advanced level" indicates postbaccalaureate. If type is listed as "homepage," multiple resources are provided by that organization.

organized by sections including What is Green Chemistry?, Design Principles, Roundtables, Research and Innovation, Students and Educators, and the Green Chemistry Institute (GCI). The first two sections cover the basics of green chemistry and green chemistry engineering principles. The Roundtables section links to several specific groups that work together around a common theme to discuss green chemistry and how to effectively "green" their disciplines. Roundtable areas include pharmaceuticals, hydraulic fracturing, formulations, chemical manufacturing, and biochemistry technology. The Research and Innovation section connects to several areas of green chemistry that can be used by professionals or students (e.g., catalysis, green chemistry and engineering metrics, solvents, and toxicology are areas listed for further exploration). The section for Students and Educators has an expansive array of links to academic programs, summer schools, workshops, activities suggested for ACS student chapters, and an education roadmap for incorporating green chemistry into the undergraduate curriculum. Finally, the last tab on the ACS Green Chemistry website links to the GCI, a nonprofit organization formed in 1997 with strategic goals for science, education, and industry. Among other activities, the

GCI hosts a yearly Green Chemistry and Engineering conference and is governed by an advisory board.

1.5.5.2 Beyond Benign

Beyond Benign equips educators with strategies and tools needed to integrate green chemistry into their classrooms from K-12 through higher education. Their website features experiments, case studies, and enrichment curriculum topics as well as community engagement opportunities for educators to discuss and share ideas. This organization also provides professional development through their workshops, webinars, and collaborative working groups. Beyond Benign is a very resourceful organization for green chemistry education and serves as the creator and host for the Green Chemistry Commitment pledged by many colleges and universities. The Green Chemistry Commitment requires both faculty and administration to pledge to incorporate principles of green chemistry into their curricula. As of March 2019, a total of 54 schools had signed the pledge *(30)*.

1.5.5.3 The United States Environmental Protection Agency

The EPA provides information about the principles of green chemistry and links to funding sources for green chemistry research and tools as well as connections to educational resources. One of the primary ways that the EPA has promoted green chemistry is through sponsoring the Green Chemistry Challenge Awards (previously the Presidential Green Chemistry Challenge Awards). These awards are traditionally made in three focus areas (1) greener synthetic pathways; (2) greener reaction conditions; and (3) design for greener chemicals. There are also two additional awards: one for small businesses and one for academia. These Green Chemistry Challenge Awards have provided the basis of case studies in green chemistry education *(36)*.

1.5.5.4 Michigan Green Chemistry Clearinghouse

The Michigan Green Chemistry Clearinghouse was developed through the Michigan Green Chemistry program in 2006 and sponsored by the state Department of Environmental Quality. This website advertises green chemistry events and activities in Michigan. The homepage of the Michigan Green Chemistry Clearinghouse highlights reference materials for the general public, education, industry, and government. The educational resources section has external links to websites featuring educational materials and provides green chemistry syllabi, lecture materials, example homework and examinations. Funding initiatives, conferences, social media links, and industry tools are also available at this website.

1.5.5.5 Center for Green Chemistry and Green Engineering at Yale University

This Center has developed green educational resources that can be utilized at various educational levels. The Molecular Design Research Network (MoDRN) at Yale has developed curricular modules for high school level chemistry, biology, and environmental sciences. These modules introduce the concepts of green chemistry and sustainable chemical design to high school students and are presented in accordance with Next Generation Science Standards. The high school modules are fully written lessons in downloadable format, and include topics of laboratory safety, understanding safety data sheets, chemical toxicity, hypothesis testing, and environmental justice. This website also provides an interactive game for undergraduate students (*The Safer Chemical Design Game*) which is focused on understanding toxicity. The game introduces students to chemical design that would minimize negative or undesirable biological impact and aims to mimic real-world constraints that students may face within an industry or research setting.

The MoDRN team have also developed 10 online introductory modules called "MoDRN: U modules" that are free to use and recommended for incorporation into an undergraduate chemistry curriculum. These modules fall under three main categories: (1) physiochemical properties; (2) principles of toxicity; and (3) physiochemical properties and toxicity in chemical design. There are learning objectives for each module as well as comprehensive assignments designed to help students make connections between interdisciplinary topics. The material is based upon work supported by the NSF Division of Chemistry and the EPA. This resource also contains green chemistry videos developed by the MoDRN team and Paul Anastas. The videos offer examples and key definitions from a green chemistry perspective divided into the areas of toxicology, solvents, energy, feedstocks, catalysis, end-of-life, accidents, and design of safer chemicals. Professional education materials including online certification programs and course offerings to assist people facing sustainability challenges in the workplace are also provided at this Center.

1.5.5.6 The University of Scranton

Faculty at the University of Scranton have been deeply involved in creating green educational materials for many years. They have contributed to writing textbooks, case studies, online learning modules and annotated bibliographies in green chemistry, as well as holding workshops and providing links to many green chemistry resources on their webpage. The University of Scranton was awarded the Camille and Henry Dreyfus Foundation Program Grant for the chemical sciences, which supported six faculty members to develop green chemistry teaching modules pertaining to specified areas of chemistry. Additional funding for their project was provided by the ACS/EPA Green Chemistry Educational Materials Development Project. The areas of chemistry that have green modules include general chemistry, organic, inorganic, biochemistry, environmental, polymer, advanced organic, chemical toxicology, and industrial chemistry. Each module contains three parts: the module web page, notes to instructors, and PowerPoint presentations. The notes to instructors suggest how the module can be incorporated into a course, and each PowerPoint presentation can be downloaded by students and instructors to take notes and be used in lectures. The modules are aimed to serve as supplemental green lessons to teach in conjunction with the required course topics.

1.5.5.7 The Institute for Green Science at Carnegie Mellon University

The Institute for Green Science led by Terry Collins at Carnegie Mellon University offers a free online course established to lead others to pursue science in a sustainable way, mainly through education about the principles of green chemistry. The course has an easy-to-follow curriculum layout of four modules, each with nine lessons. Each lesson incorporates learning tools such as games, videos, lectures, and quizzes. This resource provides a simple introductory offering to anyone interested in learning about sustainable science and green chemistry. The Institute also has resources and links to educational websites, journals, and videos that contain information about sustainability and green education for many grade levels.

1.6 Conclusion

Since the inception of green chemistry in the 1990s, many scientists and educators have accepted the challenge to consider sustainability in their practice of chemistry. Educational resources for green chemistry are numerous and will continue to be created and shared. Since the field is inherently concerned with sustainability and requires creativity in its application, its subject matter is well poised for incorporating best practices in education. As the authors of this chapter, it is our hope that we have inspired educators to investigate numerous resources to help them teach green chemistry, while staying focused on how they may incorporate each resource in a pedagogically productive and inclusive manner. A green chemistry course that incorporates essential learning outcomes, includes HIPs, and considers inclusion in its design has the most potential to be an impactful educational experience for all students.

References

1. Anastas, P. T.; Warner, J. C. *Green Chemistry: Theory and Practice;* Oxford University Press: New York, NY, 1998; p 30.
2. Morra, B.; Dicks, A. P. Recent Progress in Green Undergraduate Organic Laboratory Design. In *Green Chemistry Experiments in Undergraduate Laboratories,* Vol. 1233, Fahey, J. T., Maelia, L. E., Eds.; American Chemical Society: Washington, DC, 2016; pp 7–32.
3. Gross, E. M. Green Chemistry and Sustainability: An Undergraduate Course for Science and Nonscience Majors. *J. Chem. Educ.* **2012,** *90,* 429–431.
4. Haley, R. A.; Ringo, J. M.; Hopgood, H.; Denlinger, K. L.; Das, A.; Waddell, D. C. Graduate Student Designed and Delivered: An Upper-level Online Course for Undergraduates in Green Chemistry and Sustainability. *J. Chem. Educ.* **2018,** *95,* 560–569.

5. Kennedy, S. A. Design of a Dynamic Undergraduate Green Chemistry Course. *J. Chem. Educ.* **2015,** *93,* 645−649.

6. Manchanayakage, R. Designing and Incorporating Green Chemistry Courses at a Liberal Arts College to Increase Students' Awareness and Interdisciplinary Collaborative Work. *J. Chem. Educ.* **2013,** *90,* 1167−1171.

7. Marteel-Parrish, A. E. Toward the Greening of our Minds: A new Special Topics Course. *J. Chem. Educ.* **2007,** *84,* 245−247.

8. Prescott, S. Green Goggles: Designing and Teaching a General Chemistry Course to Nonmajors Using a Green Chemistry Approach. *J. Chem. Educ.* **2013,** *90,* 423−428.

9. American Chemical Society Committee on Professional Training. *Undergraduate Professional Education in Chemistry: ACS Guidelines and Evaluation Procedures for Bachelor's Degree Programs,* 2015. https://www.acs.org/content/dam/acsorg/about/governance/committees/training/2015-acs-guidelines-for-bachelors-degree-programs.pdf.

10. Freeman, T. M.; Anderman, L. H.; Jensen, J. M. Sense of Belonging in College Freshmen at the Classroom and Campus Levels. *J. Exp. Educ.* **2007,** *75,* 203−220.

11. Kuh, G. D. *High-impact Educational Practices: What They Are, Who Has Access to Them, and Why They Matter;* Association of American Colleges and Universities: Washington, DC, 2008.

12. Freeman, S.; Eddy, S. L.; McDonough, M.; Smith, M. K.; Okoroafor, N.; Jordt, H.; Wenderoth, M. P. Active Learning Increases Student Performance in Science, Engineering, and Mathematics. *Proc. Nat. Acad. Sci.* **2014,** *111,* 8410−8415.

13. *The LEAP Challenge.* www.aacu.org/leap.

14. Williams, D. A.; Berger, J. B.; McClendon, S. A. *Toward a Model of Inclusive Excellence and Change in Postsecondary Institutions;* Association of American Colleges and Universities: Washington, DC, 2005.

15. National Leadership Council. *College Learning for the New Global Century;* Association of American Colleges and Universities: Washington, DC, 2007. https://www.aacu.org/sites/default/files/files/LEAP/GlobalCentury_final.pdf.

16. Collins, T. Toward Sustainable Chemistry. *Science* **2001,** *291,* 48−49.

17. Rhodes, F. H. T. Sustainability: The Ultimate Liberal Art. *Chron. High Educ.* **2006,** *53.*

18. *Sustainability Curriculum in Higher Education: A Call to Action;* Association for the Advancement of Sustainability in Higher Education: Denver, CO, 2010.

19. Weissman, N. B. Sustainability & Liberal Education: Partners by Nature. *Lib. Educ.* **2012,** *98,* 4.

20. Hussar, W. J.; Bailey, T. M. *Projections of Education Statistics to 2022;* DOE, 2014.

21. *Howard Hughes Medical Institute Inclusive Excellence.* https://www.hhmi.org/developing-scientists/inclusive-excellence.

22. REALISE: REALising Inclusive Science Excellence in Biology, Chemistry, and Physics at Radford University. https://www.radford.edu/content/csat/home/realise.html.

23. del Carmen Salazar, M.; Norton, A. S.; Tuitt, F. A. Weaving Promising Practice for Inclusive Excellence Into the Higher Education Classroom. *Improv. Acad.* **2010,** *28,* 208−226.

24. Lopez, S. J.; Louis, M. C. The Principles of Strengths-based Education. *J. Coll. Charact.* **2009,** *10,* 1−8.

25. American Chemical Society Committee on Professional Training. *Green Chemistry in the Curriculum,* 2018. https://www.acs.org/content/dam/acsorg/about/governance/

committees/training/acsapproved/degreeprogram/green-chemistry-in-the-curriculum-supplement.pdf.

26. American Chemical Society: Diversity & Inclusion. https://www.acs.org/content/acs/en/membership-and-networks/acs/welcoming/diversity.html.

27. Brownell, S. E.; Tanner, K. D. Barriers to Faculty Pedagogical Change: Lack of Training, Time, Incentives, and…Tensions With Professional Identity? *CBE-Life Sci. Educ.* **2012,** *11,* 339–346.

28. Ross, K. A. *Breakthrough Strategies: Classroom-Based Practices to Support New Majority College Students;* Harvard Education Press: Cambridge, MA, 2016.

29. Manahan, S. E. *Green Chemistry and the Ten Commandments of Sustainability;* Chem-Char Research Inc. Publishers: Columbia, MO, 2011.

30. Beyond Benign: The Green Chemistry Commitment. https://www.beyondbenign.org/he-green-chemistry-commitment.

31. Community Resources for Science. http://www.crscience.org/pdf/Marshmallow_toothpick_challenge.pdf.

32. Inclusive Teaching Toolkit. https://www.monash.edu/library/inclusive-teaching.

33. Lancaster, M. *Green Chemistry: An Introductory Text,* 3rd ed.; Royal Society of Chemistry: Cambridge, 2016.

34. Török, B., Dransfield, T., Eds. *Green Chemistry: An Inclusive Approach;* Elsevier: Amsterdam, 2018.

35. Ryan, M. A., Tinnesand, M., Eds. *Introduction to Green Chemistry: Instructional Activities for Introductory Chemistry;* American Chemical Society: Washington, DC, 2002.

36. Cann, M. C.; Connelly, M. E. *Real-World Cases in Green Chemistry;* American Chemical Society: Washington, DC, 2000.

37. Environmental Protection Agency: Green Chemistry Challenge Winners. https://www.epa.gov/greenchemistry/green-chemistry-challenge-winners.

38. Kirchhoff, M.; Ryan, M. A. *Greener Approaches to Undergraduate Chemistry Experiments;* American Chemical Society: Washington, DC, 2002.

39. Doxsee, K. M.; Hutchison, J. E. *Green Organic Chemistry: Strategies, Tools, and Laboratory Experiments;* Brooks/Cole Publishing Company: London, 2004.

40. Roesky, H. W., Kennepohl, D. K., Eds. *Experiments in Green and Sustainable Chemistry;* Wiley-VCH: Weinheim, Germany, 2009.

41. Perosa, A.; Zecchini, F. *Methods and Reagents for Green Chemistry: An Introduction;* John Wiley & Sons: Hoboken, NJ, 2007.

Incorporating elements of green and sustainable chemistry in general chemistry via systems thinking

2

Thomas Holme, PhD

Morrill Professor, Department of Chemistry, Iowa State University, Ames, IA, United States

2.1 Introduction

The role of introductory college science courses, including general chemistry, has long been defined in terms of fundamental concepts that students need to know to proceed to subsequent courses. While this core component is unlikely to be entirely replaced, recent developments in understanding how people learn *(1,2)* suggest that explicitly helping students to transfer fundamental knowledge to broader topics is a key aspect for teaching and learning in Science, Technology, Engineering and Math (STEM). The challenge becomes, therefore, how to infuse a course, such as general chemistry, with enough connections to broader themes and skills to allow students to transfer knowledge while maintaining appropriate attention on fundamental concepts.

The approach suggested here is to have an organizing theme that connects fundamental concepts in chemistry to the framework of planetary boundaries *(3–5)* and uses concepts associated with systems thinking *(6)*. This choice represents an excellent method for accomplishing the balance between core concepts and their application to sustainability, in terms of earth and societal systems. Systems thinking has emerged as an important component of recent efforts to enhance STEM education. In particular, at the K-12 level in the United States, the Next Generation Science Standards (NGSS) include systems thinking and modeling as a cross-cutting concept *(7)*. There are a number of definitions for the application of systems thinking in educational settings *(8–10)*. These definitions are similar but not clearly equivalent. As may be inferred from the existence of multiple definitions of systems thinking in education, there has been significant effort applied to the incorporation of these ideas in various disciplines, but less so within chemistry. For example, a recent review emphasizes educational developments in biology, engineering, environmental science, and geosciences in enumerating systems thinking work *(11)*. Despite this emphasis, however, there are topics where efforts in other fields clearly touch on topics in general chemistry, such as the water cycle and water usage *(8,12)*.

Thus, for the purpose of this chapter, the key aspects of systems thinking will be considered as the ability to not only identify and understand components of a system but also to recognize the framework within which they are dynamically related. This capacity is vital to be able to identify emergent properties that arise within systems. Finally, it is important to recognize that there is an additional challenge associated with providing meaningful assessment instruments, in addition to pedagogical interventions designed to enhance students' systems thinking *(13)*.

To advance the discussion of incorporation of systems thinking attributes in general chemistry, it is helpful to consider those described by Assaraf and Orion *(8)* in their description of systems thinking in the context of geosciences education. These authors have enumerated a set of eight traits that characterize systems thinking, including (1) the ability to identify the components of a system and the processes within the system; (2) the ability to identify dynamic relationships among the system components; (3) the ability to identify dynamic relationships within the system; (4) the ability to organize the systems' components and processes within a framework of relationships; (5) the ability to understand the cyclic nature of many systems; (6) the ability to make generalizations; (7) the ability to understand the hidden dimensions of the system; and (8) the ability to think temporally, including retrospection and prediction. These attributes can serve as a foundation by which core chemistry knowledge is connected to issues of sustainability, and thus allow the incorporation of both systems thinking and sustainability in general chemistry.

Another important aspect of how systems thinking can be implemented into general chemistry arises by using a systems framework to question the appropriateness of boundaries that are considered for a given question or problem. Thus, in addition to other attributes, systems thinking provides impetus to ask the question: are we considering the right boundary for this process? *(14)*. This is a powerful way to shape student attitudes toward core chemistry knowledge from a reductionist view toward a more expansive one that includes applications of that core knowledge *(15)*. One example is the idea that chemicals may have both benefits and hazards, and those aspects require consideration of the boundaries that are defining the scope of a given system *(16)*. A second example is the concept that multiple cycles of development often arise as chemists seek solutions to problems. Issues like chlorofluorocarbon (CFC) refrigerants being replaced by hydrochlorofluorocarbons (HCFCs) which were subsequently replaced by hydrofluorocarbons (HFCs), or the invention of antibiotics and the more recent rise in antibiotic resistance are two examples that have been suggested *(17)*.

Green chemistry also has substantial overlap with the types of reasoning that make up the central aspects of systems thinking. For example, the ability to conduct life cycle assessment *(18)* in the creation of products such as textiles *(19)* includes aspects of systems thinking *(20)*. Most reports of the incorporation of green chemistry concepts in introductory chemistry, however, have placed their focus on the laboratory, often with an emphasis on chemical waste reduction as a key driver

(21−23). Nonetheless, the design principles of green chemistry are ideal vehicles to help students ask the question of what boundary should be used in thinking about a chemical process. In particular, thinking about the origin or synthesis of reactant materials often provides an entryway into a more systems thinking approach to understanding fundamental chemistry ideas.

Beyond green chemistry and the design principles for which it is best known, the concept of infusing sustainability more broadly into a general chemistry curriculum also shows appeal to many students. A challenge here lies in the often-noted tendency to require multidisciplinary approaches to sustainability courses *(24)*. Even so, with its historical moniker as the "central science," chemistry courses are well placed to incorporate connections to other disciplines by incorporating explanations of chemical issues within rich contexts *(17,25,26)*. While much work has been put forward within the framework of rich contexts, the ability to use an overarching organizational principle for viewing sustainability adds a new and helpful dimension to this style of educational intervention in chemistry classes.

2.1.1 Planetary boundaries

The key concept that provides the overall framework discussed here and allows for the connection of systems thinking to sustainability is planetary boundaries *(3−5)*. While concerns have been noted regarding this concept *(27)*, it provides an important route for thinking of both components and larger systems. This framework is particularly helpful in the context of applying systems thinking in general chemistry. The central concept of planetary boundaries is to take available estimates of global scale issues and assess whether or not a sustainability "tipping point" has been surpassed due to human activities. Fig. 2.1 provides a representation of planetary boundaries that is useful in emphasizing the role of chemistry, including concepts taught in general chemistry, in understanding planetary boundaries.

In this figure, based on the most recent estimates of the planetary boundaries *(4)*, the purple horizontal line represents the putative tipping point for the characteristic, and those systems that are below it and in red tones have already surpassed the probable tipping point. Using this representation consistently in class to connect various foundational chemistry topics to the idea of sustainability carries the memorable tagline for students, "Does this chemistry help us understand whether we are above the bar or not?" In some cases (specifically chemical pollutants and atmospheric aerosols), insufficient data are available to make an estimate. Finally, characteristics that remain above the line and in shades of green have not yet reached an estimated tipping point. Original reports *(3−5)* included numerical estimates as percentages toward the tipping point, but as an organizational framework such detail is not required, and could be counterproductive in the context of general chemistry where some aspects of the planetary boundaries issues are beyond the scope of the science content of the course. One reason this framework works so well for infusing elements of systems thinking into general chemistry lies in the fact that most of the areas of key concerns have significant chemical aspects associated with them. For

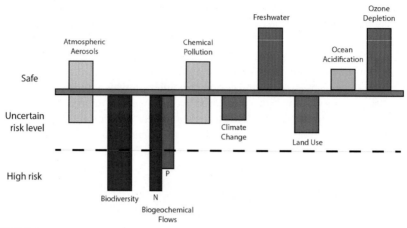

FIGURE 2.1

Depiction of the categories considered in the Planetary Boundaries framework. Categories below the purple horizontal line have crossed the estimated tipping point, while those above the line have not. Categories in gray have insufficient data to determine whether or not a tipping point has been crossed. For "Biogeochemical Flows", *N*, nitrogen and *P*, phosphorus.

*Adapted from Steffen, W.; Richardson, K.; Rockström, J.; Cornell, S. E.; Fetzer, I.; Bennett, E. M.; Biggs, R.; Carpenter, S. R.; de Vries, W.; de Wit, C. A.; Folke, C.; Gerten, D.; Heinke, J.; Mace, G. M.; Persson, L. M.; Ramanathan, V.; Reyers, B.; Sörlin, S. Planetary Boundaries: Guiding Human Development on a Changing Planet. Science **2015**, 347, 1259855.*

example, both areas where estimates cannot yet be made involve chemistry, and this fact emphasizes how furthering knowledge and research in chemistry is important for devising appropriate approaches to issues related to sustainability. Further, the issues of biogeochemical flows and climate change, which are estimated to have surpassed a tipping point, also clearly involve chemistry as both a source for the current concern, and quite likely as a component of any solutions that can be developed to address the problems. This chapter will describe several examples of how this is done in one particular general chemistry course.

2.2 Considerations of course design

When devising interventions that allow instruction to highlight issues related to sustainability, it is vitally important to acknowledge and account for constraints associated with the larger curricular context of the course. This idea is particularly true for a "gateway course" such as general chemistry that is required for a large number of college degrees in the sciences. Thus, curricular innovations undertaken in this course are not that similar to what might be accomplished, for example, in a dedicated, interdisciplinary course *(23)*. There are undoubtedly many ways that this type of consideration can be accounted for, but the approach described here emphasizes incorporating ideas related to sustainability in the chemistry course objectives *(28,29)*. Communication with students that articulates outcomes they perceive they can achieve represents an important aspect of successful teaching and course improvement *(30)*. Therefore, while the core content knowledge to be covered in a general chemistry course is vital to maintain and emphasize, adding systems thinking objectives must also be clearly communicated and connected to the larger goals for student learning in the course.

In chemistry education over the past several years, efforts led by the American Chemical Society Examinations Institute (ACS-EI) have included mapping the 4-year curriculum in terms of anchoring concepts or "big ideas" *(31−33)*. This tool, called the Anchoring Concepts Content Map (ACCM), has also been used to identify gaps in the traditional curriculum, particularly with respect to conceptual knowledge with applications in sciences outside of chemistry *(34)*. The results of this process for only the first semester of general chemistry are depicted in Fig. 2.2. ACS examinations are produced by committees of instructors from across the United States and tend to reflect content coverage more succinctly than textbooks, which routinely include more topics than any instructor teaches. Therefore, this histogram represents a reasonable estimate of what topics are considered important in the curriculum, and it reveals information both in terms of topics that appear often, as well as those that are seldom tested. The incorporation of sustainability content and systems thinking needs to be accomplished without dramatically changing this content coverage. Ideally, therefore, incorporation of these newer ideas would be achieved within areas that already provide important focus for the course.

Given this premise, areas that have been emphasized over the past 20 years in the first semester course include atomic structure, Valence Shell Electron Pair Repulsion to assign molecular shape, relative strength of intermolecular forces (IMFs), physical properties and gases, balancing chemical equations and categorizing reactions, stoichiometry and limiting reactants, and energy and thermochemistry. While connecting with these key topics represents one way to address potential concerns about taking time to include sustainability and systems thinking, it is also worth noting areas that have not been part of content coverage over the years. For example, as noted previously *(34)*, despite having a large clientele interested in the life sciences, general chemistry instruction has not been very adept at including noncovalent forces within larger biomolecular systems in the coverage of IMF. There are also important ideas that have garnered little enough attention that they do not even appear in the ACCM itself. One noteworthy example of this is that introductory principles of toxicology, including toxicity and exposure, are not present in either version of the General Chemistry ACCM *(31,32)*, despite the importance of these concepts in a number of fields and in the manner in which chemistry affects society. Indeed, the role of toxicity is incorporated into the first day of class in this model, as a core component of understanding the risk/benefit analysis of chemical systems.

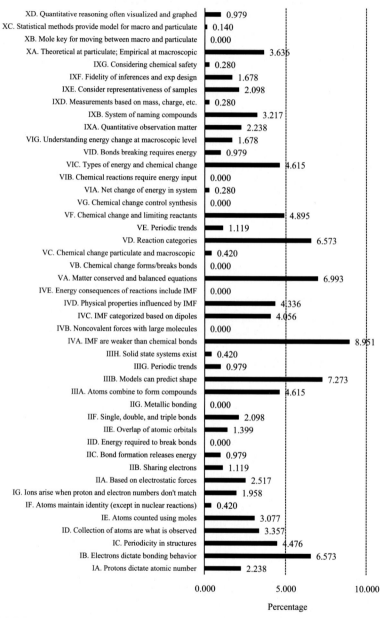

**Percentage of Items on ACS 1st Term
General Chemistry Examination**

XD. Quantitative reasoning often visualized and graphed	0.979
XC. Statistical methods provide model for macro and particulate	0.140
XB. Mole key for moving between macro and particulate	0.000
XA. Theoretical at particulate; Empirical at macroscopic	3.636
IXG. Considering chemical safety	0.280
IXF. Fidelity of inferences and exp design	1.678
IXE. Consider representativeness of samples	2.098
IXD. Measurements based on mass, charge, etc.	0.280
IXB. System of naming compounds	3.217
IXA. Quantitative observation matter	2.238
VIG. Understanding energy change at macroscopic level	1.678
VID. Bonds breaking requires energy	0.979
VIC. Types of energy and chemical change	4.615
VIB. Chemical reactions require energy input	0.000
VIA. Net change of energy in system	0.280
VG. Chemical change control synthesis	0.000
VF. Chemical change and limiting reactants	4.895
VE. Periodic trends	1.119
VD. Reaction categories	6.573
VC. Chemical change particulate and macroscopic	0.420
VB. Chemical change forms/breaks bonds	0.000
VA. Matter conserved and balanced equations	6.993
IVE. Energy consequences of reactions include IMF	0.000
IVD. Physical properties influenced by IMF	4.336
IVC. IMF categorized based on dipoles	4.056
IVB. Noncovalent forces with large molecules	0.000
IVA. IMF are weaker than chemical bonds	8.951
IIIH. Solid state systems exist	0.420
IIIG. Periodic trends	0.979
IIIB. Models can predict shape	7.273
IIIA. Atoms combine to form compounds	4.615
IIG. Metallic bonding	0.000
IIF. Single, double, and triple bonds	2.098
IIE. Overlap of atomic orbitals	1.399
IID. Energy required to break bonds	0.000
IIC. Bond formation releases energy	0.979
IIB. Sharing electrons	1.119
IIA. Based on electrostatic forces	2.517
IG. Ions arise when proton and electron numbers don't match	1.958
IF. Atoms maintain identity (except in nuclear reactions)	0.420
IE. Atoms counted using moles	3.077
ID. Collection of atoms are what is observed	3.357
IC. Periodicity in structures	4.476
IB. Electrons dictate bonding behavior	6.573
IA. Protons dictate atomic number	2.238

0.000   5.000   10.000

Percentage

FIGURE 2.2

Percentage coverage of "enduring understandings" of 715 items in First Term General Chemistry produced by the ACS Examinations Institute spanning 20 years of tests.

From the ACCM.

Box 2.1 Content curriculum of first semester general chemistry

1. Introduction, particles and matter
   a. includes units of measure
2. Atoms, ions and molecules
   a. includes initial introduction to the periodic table, nucleosynthesis, and isotopes
3. Stoichiometry
4. Aqueous reactions/water chemistry
5. Thermochemistry
6. Gases and their properties
7. Quantum model of atoms and atomic structure
   a. includes periodicity
8. Chemical bonding
9. Molecular geometry
10. Intermolecular forces
    a. includes liquids and physical properties of liquids

In addition to this overview of content, it is also important to note that the text-book is often the primary artifact that guides curriculum. Most general chemistry courses, particularly those with multiple sections taught by different instructors, tend to follow the chapter organization of textbooks. The first semester course where systems thinking components were added follows this pattern as well, with the chronological order of topics as depicted in Box 2.1.

This general ordering needs to be maintained in a multiple-section teaching environment, so our efforts to infuse sustainability are accomplished within this broad outline. Because the overall outline is maintained, there are inherent time constraints on the amount of material that can be introduced to support student awareness of systems thinking.

2.3 Incorporation of sustainability and systems thinking

The commitment of time to coordinating sustainability concepts in general chemistry begins on the first day of the course. Noting the explicit learning outcomes that connect foundational concepts to applications of earth and societal systems represents a vital part of the introductory activities. This includes the initial presentation of the concept of planetary boundaries as depicted earlier in Fig. 2.1. The first introduction of new, broader context ideas that can be related to systems thinking arises when the periodic table is discussed in the second content area. At this time, we include the concept of ranges of atomic weights present in the most recent IUPAC-approved periodic table *(35)*. At this early point, the connection to climate change and evidence related to it is only mentioned and connected to the planetary boundaries construct, without any details. Rather, it is noted that later in the course, the idea that isotopic abundance can vary in nature leads to an ability to estimate "deep time" in ice cores will be an important concept in understanding the evidence about climate. Similarly, in the third topic area (stoichiometry), we include an emphasis on the reaction of nitrogen and hydrogen to form ammonia. This reaction is used as the key example for the description of particulate nature of matter (PNOM) diagrams *(36)* within stoichiometry, for example. These early additions are essentially examples that both advance the core content and foreshadow the role that foundational chemistry knowledge plays in understanding planetary boundaries.

2.3.1 Water chemistry and biogeochemical flows

The first time when issues related to planetary boundaries arises in a way that can strongly motivate the chemistry covered arises in the fourth topic area: water chemistry. This instance of incorporating systems thinking has been described in greater detail elsewhere *(37)*, but it centers on biogeochemical flows, particularly of nitrogen. Because Iowa State is located in an agricultural region of the United States, the large-scale application of nitrogen-based fertilizer represents a regional issue in addition to being part of a larger, planetary boundary content area. These characteristics, however, play an important role in implementing the concept of exploring boundaries as a way to introduce systems thinking concepts. In this case, the foundational chemistry concept is solubility and solubility rules. Solubility rules are routinely a source of consternation for students, in part because they seem to arise from arbitrary lines drawn along a continuum of solubility *(38)* and partly because they seem rather compact and used for a small portion of the curriculum. The primary way students are accustomed to interacting with solubility rules lies in assessments that ask "which pair of solutions, when mixed, will form a precipitate?" so that the rules are pointedly oriented toward a laboratory activity with little relevance to students outside of that activity.

Nonetheless, perhaps the most commonly remembered solubility rule is that all nitrates are soluble, and this allows the introduction of how nitrates, applied in fertilizers, enter the watershed. Because of this, local issues related to drinking water standards provide the first "expanded boundary" with solubility getting "out of the laboratory" and making a difference in how municipalities produce safe drinking water. Because this aspect of nitrates in rivers also incorporates standards for drinking water based on at-risk populations, specifically infants in this case, it also requires the further exploration of ideas of toxicology *(39)*. Therefore, taking a modest amount of time to situate solubility in a regional level environmental concern serves to highlight both systems thinking and aspects of green chemistry. Importantly, it also allows multiple connections to planetary boundaries if the boundaries being considered expand beyond the local, agricultural region. Specifically, the role of Midwestern agricultural practices on the creation of the hypoxic "dead zone" in the Gulf of Mexico becomes important if we ask, "what if thinking about nitrates in Iowa rivers is not a large enough boundary to consider?"

This expanded boundary question invokes several connections to the planetary boundaries concepts depicted in Fig. 2.1. Most recognizably, the connections to fresh water usage, and the importance of water treatment chemistries is readily noted. Perhaps less obvious, but equally important is the concept of land use changes. This idea is approached by noting the differences between historical land use patterns of tall-grass prairies in Iowa compared to modern row-crop plants. Finally, the chemical consequences of (1) nitrate loading into rivers; (2) the flow of those nutrients to distant regions; and (3) their effects when they arrive there show how larger scale systems thinking concepts are important. Just as important, however, is the ability to make the argument that beginning to understand the larger scale system is dependent on understanding the components of those systems: in this case the foundational chemistry concept of the role of solubility of nitrates in water.

Finally, one additional aspect of solution chemistry is introduced with the idea of foreshadowing in mind once again. Thus, the use of Beer's law to determine concentration of solutions that include absorbing solutes is covered. The ability to have the concept of $A = \varepsilon bc$ and noting that the extinction coefficient, ε, is dependent on the frequency of light being applied will be used again when climate change is discussed.

The traditional ordering of chemistry topics, in terms of textbook chapters, may diverge somewhat at this point in the course based on the specific book being used, but in our situation the next topic covered is thermochemistry. Because of the importance of the Haber−Bosch process and the role of energy in the fixing of atmospheric nitrogen, the biogeochemical flow of nitrogen can be further accentuated immediately following the treatment of water chemistry. This connection reinforces the systems thinking aspects by "expanding the boundary" in a new direction, one that is about the origin of the chemicals, rather than their fate in the environment. These first examples of incorporating aspects of chemistry related to the biogeochemistry of nitrogen reflect an implementation strategy where the new content changes little about the traditional presentation of the content curriculum. In this specific case, the order of topics in water chemistry and thermochemistry is largely unchanged, but examples are pulled from the rich context of nitrogen flow in the environment. This allows connections to planetary boundaries and sustainability, without taxing instructional resources.

2.3.2 Gases and climate change

When incorporating concepts related to climate change, however, both content additions and changes to customary ordering arise. These changes affect the coverage of gases and their properties the most. Where traditional coverage of this material emphasizes the sameness of ideal gases, such that a single equation is usable for many aspects of any such gas, a meaningful connection of atmospheric gases to climate change requires a different approach. In addition, for a course at the university level, gas law applications and calculations seem quite familiar to many students from prior exposure to them in high school. This familiarity can lead to a decrease in attention and motivation for a large fraction of the students in a course because they perceive their prior knowledge to be adequate to meet expectations for that particular content.

Therefore, in the approach that connects to climate change, the chapter on gases begins with a direct connection to prior ideas in the course, from the just-covered chapter on energy. As such, the connection between combustion, energy, and the production of CO_2 is noted, and the question of what boundary needs to be considered leads to the temporal nature of the evidence. Specifically, it demands an explanation of how science knows about temperature and atmospheric composition over long time periods. This question leads immediately to the concept of "deep time" *(40)* and ways to infer information about climate hundreds of thousands of years ago. Establishing the ability to measure concentrations of CO_2 over time is readily introduced in terms of bubbles trapped in the ice, but the measurement of temperature requires the return to the idea of isotopes and atomic weight. The difference between the movement of ^{16}O water and ^{18}O water because of the difference in mass fits well in the flow of the course because it is related to the concept of the Boltzmann distribution, which is part of the coverage in the chapter on energy. That these ideas also connect to atmospheric dynamics, an important aspect of systems thinking, is essentially a bonus for making the connections to systems thinking skills. Having used this concept to introduce the movement of gas molecules, the approach taken to cover this chapter is also changed. The content coverage begins with gas kinetic theory rather than the more traditional approach of describing historical observations of gas laws. This change has the effect of catching the attention of students who might otherwise be inclined to "coast" through previously seen gas law problems.

Once gas kinetic theory has been established, it is used to explain the historical gas laws, as is commonly done, but the question of boundaries is quickly introduced again. The hook used is to ask the boundary question in terms of what is gained and what is lost by treating all gases alike, which is the power of the ideal gas law. Nonetheless, it does not answer the key question of why CO_2 is a greenhouse gas, while neither N_2 nor O_2 are. This is the moment when we are able to connect most strongly to climate change as an area where a tipping point has been reached, by reintroducing Beer's law, which had been included in the earlier work on solutions. In this context, the coverage focuses on both concentration and the extinction coefficient.

The concentration discussion emphasizes the changes in CO_2 in the atmosphere, including the concepts from the initial deep time discussion, but also invokes the isotopic abundance concepts to establish the anthropogenic nature of the added CO_2 since the dawn of the industrial revolution *(26)*. The extinction coefficient discussion then centers on what renders gas molecules capable of absorbing IR radiation. The relative simplicity of the molecules involved allows this conversation to occur prior to a detailed discussion of molecular shape, which is later in the course structure. Ultimately, the role of symmetry of different molecular vibrations is introduced as the key factor that makes the molecules different. This approach does require students to come to terms with the idea that molecules vibrate. As has been noted in a previous study, preinstruction surveys found only one-third of the students in general chemistry classes are aware that molecules vibrate *(41)*. After instruction based on the need to understand the role of molecular vibration in identifying greenhouse gases, however, correct student conceptual understanding of greenhouse gas vibrational motion exceeded 80%. What has been added since these earlier studies is the idea of linking the concept of molecular vibration to systems thinking by asking the question about the correct boundary. In this case, the expanded boundary is whether or not the ideal gas model, where the identity of the gas is not important, is enough to explain all properties of atmospheric gases that are important to the question of climate change. Ultimately, N_2, O_2, and CO_2 all behave as ideal gases in the atmosphere, but more than the ideal gas model is needed to be able to understand key, larger scale issues.

These examples are designed to highlight how green and sustainable chemistry concepts can be integrated in a coherent way as part of the general chemistry course. The coherence itself is provided in two ways. As noted in Fig. 2.3, we first connect

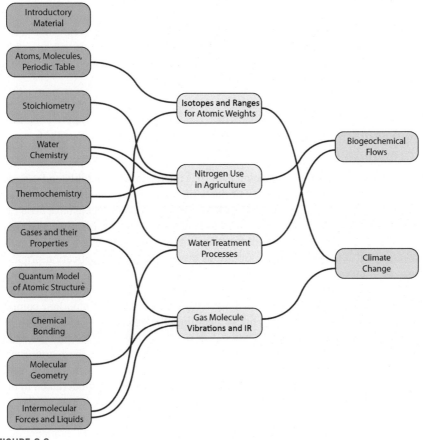

FIGURE 2.3

Mapping of planetary boundaries (*green boxes*) to rich context topics (*yellow boxes*) that map to traditional, chapter-based content organization for the first semester of general chemistry (*blue boxes*).

all of the content coverage modifications to the idea of planetary boundaries. Returning to the same construct, and focusing on only two aspects of it, allows students to build depth of understanding through a narrative of how chemistry affects broader scientific issues, rather than acting as a compartmentalized knowledge base. The other aspect of the coherence is the process by which we routinely engage with the broader context. Specifically, the repeated questioning of whether or not we need to consider broader boundaries for the science as we cover it becomes a familiar process, one that accentuates the need to make the broader connections. Finally, it is worth noting that the connections shown in Fig. 2.3 are not the only possible ones. For example, it would be reasonable to connect bonding to gas molecule vibrations through the relationships between vibrational energy and bond strength, but in the current course, this idea has not been implemented.

It is also worthwhile to note that Fig. 2.3 is limited to first semester general chemistry content and possible connections to planetary boundaries and systems thinking. Work is underway to incorporate this strategy in the second semester of the general chemistry course, where ocean acidification, for example, is a key planetary boundary that connects readily to the content on acid/base equilibrium that is customarily covered then.

2.4 Evaluating student learning of new and connected materials

While the implementation of teaching materials related to sustainability and systems thinking represents the core activity of the work reported here, it remains important to assess whether student learning arises from this intervention. There have been two key methods by which student learning has been ascertained in this project. Firstly, when an appropriate concept inventory is available, it has been used *(41)*. Secondly, it is important for students to see questions about these more application-oriented topics on midterm examinations, or the assessments will not signal that the material is important. As such, the inclusion of items that query student understanding of these systems thinking—related applications becomes an important component of generating student buy-in to learning the material. Thus, while this evaluation of the teaching intervention has no control group (all students were exposed to the material and tested on it), the general impact of the learning can be put into the context of overall student performance in the chemistry course itself. There are two forms of evidence from assessments that have been undertaken in this teaching intervention: measuring conceptual understanding and direct testing of content knowledge in course examinations.

The conceptual learning of climate change content acquired by students uses measures that have been reported previously as part of the description of the concept inventory instruments used to accomplish this assessment *(41)*. The most important aspect of this prior work in terms of establishing efficacy of incorporating content related to planetary boundaries is the overall change in conceptual understanding in pre-/posttesting of climate change–related concepts. For a sample of 128 students, the average percentage correct in the pretest on the climate science concept inventory was 42% while the posttest was 77%. Thus, even with a rather modest level of instructional time dedicated to climate science–related materials in the gas laws chapter, the gains in conceptual understanding are noteworthy.

For students to value the added content, assessments must include items related to that content. In large courses, assessment routinely requires the use of multiple choice questions, and in the case of incorporating sustainability, items that test the concept should have psychometric parameters in line with other content in the course. This premise is confirmed for items that incorporate these themes as shown in Table 2.1.

Looking at the information in Table 2.1, we see that three out of eight items (3, 4, and 8) have percentages correct quite close to the average, three items (1, 5, and 7) have percentages correct that are notably higher than the average, and two items (2 and 6) have percentages correct considerably lower than the average. In addition, the point biserial correlation values for all eight items are positive (meaning students with higher test scores are more likely to answer correctly). There are many arguments that have been advanced to establish guidelines for what constitutes a "good" point biserial value. Values above 0.20 have been suggested as a standard that merits this distinction *(42)*, and all of these items that test connections to planetary boundaries in the course exceed this standard. This is true even for items with very low or very high percentage correct values, which inherently have lower point biserial values. This evidence does not prove that testing of concepts related to systems approaches to incorporate sustainability is inherently strong psychometrically, but rather that it is possible to write and incorporate such items in general chemistry tests and maintain good assessment characteristics.

Table 2.1 Comparison of systems/sustainability items and item averages.

Item topic	Percentage correct (%)	Point biserial
1. Cause of lead contamination in Flint, MI	84	0.27
2. Soil biome and redox chemistry of nitrates	17	0.24
3. Large components of the natural carbon cycle	65	0.47
4. Small components of the natural carbon cycle	68	0.26
5. Deep time and CO_2 concentrations	75	0.30
6. Deep time and temperature	36	0.26
7. Vibrational motion of gases and IR radiation	84	0.36
8. Relationship between IR absorption and greenhouse gases	66	0.48
Average of all multiple-choice items	*70*	*0.42*

2.5 Discussion and conclusions

The premise of the intervention presented here is to make the argument that important curricular advances can be made while largely maintaining the traditional content coverage of general chemistry. By carefully choosing a small number of content areas related to planetary boundaries, students in first-year college chemistry courses can be encouraged to engage in several beneficial learning opportunities. Firstly, by routinely asking the question "am I using the correct boundary for understanding the role of this science?" students can be induced to move beyond compartmentalized strategies for science topics, such as those often found in chemistry. Secondly, the concept of learning more complex systems by first considering foundational level components then connecting them to larger issues represents a style of scientific thinking that needs to be emphasized as people learn chemistry and its role in understanding the world around us. Thirdly, the amount of time that needs to be used for this type of instruction, while nonzero, is not so large as to prevent a strong level of fundamental understanding of topics in the foundational courses. Finally, even though systems thinking and issues of sustainability are inherently complex topics, it is possible to both teach and assess these concepts in a meaningful way, even at the introductory level of college chemistry.

The examples presented here were clearly chosen with several aspects in mind. Firstly, the connection to planetary boundaries provides a consistent, large context to which many students can relate. Secondly, there are several science (and specifically chemistry) components encompassed within planetary boundaries that mean an individual instructor can find content within the framework with which they are comfortable. This makes the overall strategy adaptable to the skills and capacity of many instructors. The examples provided here merely illustrate one path for such teaching strategies. Thirdly, assessment efforts carried out with this particular implementation suggest that it is possible to introduce content in this way in a meaningful and measurable way in the general chemistry curriculum. Finally, the introduction of concepts related to sustainability and connecting them to green chemistry principles provides not only useful content to those for whom general chemistry is their last class. It also allows an earlier time in the curriculum to introduce these concepts and paves the way for more inclusion of green chemistry in its more common curricular location of organic chemistry.

References

1. National Research Council. *How People Learn: Brain, Mind, Experience, and School,* Expanded Ed.; National Academies Press: Washington, DC, 2000.
2. National Academies of Sciences, Engineering, and Medicine. *How People Learn II: Learners, Contexts, and Cultures;* National Academies Press: Washington, DC, 2018.
3. Rockström, J.; Steffen, W.; Noone, K.; Persson, Å.; Chapin, F. S., III; Lambin, E.; Lenton, T. M.; Scheffer, M.; Folke, C.; Schellnhuber, H.; Nykvist, B.; De Wit, C. A.;

Hughes, T.; van der Leeuw, S.; Rodhe, H.; Sörlin, S.; Snyder, P. K.; Costanza, R.; Svedin, U.; Falkenmark, M.; Karlberg, L.; Corell, R. W.; Fabry, V. J.; Hansen, J.; Walker, B. H.; Liverman, D.; Richardson, K.; Crutzen, C.; Foley, J. A Safe Operating Space for Humanity. *Nature* **2009**, *461,* 472−475.

4. Steffen, W.; Richardson, K.; Rockström, J.; Cornell, S. E.; Fetzer, I.; Bennett, E. M.; Biggs, R.; Carpenter, S. R.; de Vries, W.; de Wit, C. A.; Folke, C.; Gerten, D.; Heinke, J.; Mace, G. M.; Persson, L. M.; Ramanathan, V.; Reyers, B.; Sörlin, S. Planetary Boundaries: Guiding Human Development on a Changing Planet. *Science* **2015**, *347,* 1259855.

5. O'Neill, D. W.; Fanning, A. L.; Lamb, W. F.; Seinberger, J. K. A Good Life for All Within Planetary Boundaries. *Nat. Sustain.* **2018**, *1,* 88−95.

6. Mahaffy, P. G.; Krief, A.; Hopf, H.; Matlin, S. A. Reorienting Chemistry Education Through Systems Thinking. *Nat. Rev. Chem.* **2018**, *2,* 1−3.

7. National Research Council. *A Framework for K-12 Science Education: Practices, Cross-cutting Concepts, and Core Ideas;* National Academies Press: Washington, DC, 2012.

8. Assaraf, O. B.-Z.; Orion, N. Development of System Thinking Skills in the Context of Earth System Education. *J. Res. Sci. Teach.* **2005**, *42,* 518−560.

9. Arnold, R. D.; Wade, J. P. A Definition of Systems Thinking: A Systems Approach. *Procedia Comput. Sci.* **2015**, *44,* 669−678.

10. Verhoeff, R. P.; Boersma, K. T.; Waarlo, A. J. Multiple Representations in Modeling Strategies for the Development of Systems Thinking in Biology Education. In *Multiple Representations in Biological Education;* Treagust, D. F., Tsui, C.-Y., Eds.; Springer: The Netherlands, 2013; pp 331−348.

11. Yoon, S. A.; Goh, S.-E.; Park, M. Teaching and Learning About Complex Systems in K-12 Science Education: A Review of Empirical Studies 1995−2015. *Rev. Educ. Res.* **2018**, *88,* 285−325.

12. Lee, T. D.; Jones, M. G.; Chesnutt, K. Teaching Systems Thinking in the Context of the Water Cycle. *Res. Sci. Educ.* **2019**, *49,* 137−172.

13. Booth Sweeney, L.; Sterman, J. D. Bathtub Dynamics: Initial Results of a Systems Thinking Inventory. *Syst. Dyn. Rev.* **2000**, *16,* 249−286.

14. Boardman, J.; Sauser, B.; John, L.; Edson, R. The Conceptagon: A Framework for Systems Thinking and Systems Practice. In *Proceedings of the 2009 IEEE International Conference on Systems, Man, and Cybernetics;* 2009. San Antonio, TX.

15. Caulfield, C. W.; Maj, S. P. A Case for Systems Thinking and System Dynamics. In *Proceedings of the 2001 IEEE International Conference on Systems, Man, and Cybernetics;* 2001. Tucson, AZ.

16. Holme, T. A.; Hutchison, J. E. A Central Learning Outcome for the Central Science. *J. Chem. Educ.* **2018**, *95,* 499−501.

17. Matlin, S. A.; Mehta, G.; Hopf, H.; Krief, A. One-world Chemistry and Systems Thinking. *Nat. Chem.* **2016**, *8,* 393−398.

18. Finnveden, G.; Hauschild, M. Z.; Ekvall, T.; Guinee, J.; Reinout, H.; Hellweg, S.; Koehler, A.; Pennington, D.; Suh, S. Recent Development in Life Cycle Assessment. *J. Env. Manage.* **2009**, *91,* 1−21.

19. Moore, S. B.; Ausley, L. W. Systems Thinking and Green Chemistry in the Textile Industry: Concepts, Technologies and Benefits. *J. Clean. Prod.* **2004**, *12,* 585−601.

20. Juntunen, M.; Aksela, M. Improving Students' Argumentation Skills Through a Product Life-cycle Analysis Project in Chemistry Education. *Chem. Educ. Res. Pract.* **2014**, *15,* 639−649.

21. Klingshirn, M. A.; Spessard, G. O. Integrating Green Chemistry Into the Introductory Chemistry Curriculum. In *Green Chemistry Education;* Anastas, P. T., Levy, I. J., Parent, K. E., Eds.; American Chemical Society: Washington, DC, 2009; pp 79−92.

22. Cacciatore, K. L.; Sevian, H. Teaching Lab Report Writing Through Inquiry: A Green Chemistry Stoichiometry Experiment for General Chemistry. *J. Chem. Educ.* **2006,** *83,* 1039−1041.

23. McCarthy, S. M.; Gordon-Wylie, S. W. A Greener Approach to Measuring Colligative Properties. *J. Chem. Educ.* **2005,** *82,* 116−119.

24. Remington-Doucette, S. M.; Hiller Connell, K. Y.; Cosette, M.; Armstrong, C. M.; Musgrove, S. L. Assessing Sustainability Education in a Transdisciplinary Undergraduate Course Focused on Real−world Problem Solving: A Case for Disciplinary Grounding. *Int. J. Sustain. High. Educ.* **2013,** *14,* 404−433.

25. Bulte, A. M.; Westbroek, H. B.; de Jong, O.; Pilot, A. A Research Approach to Designing Chemistry Education Using Authentic Practices as Contexts. *Int. J. Sci. Educ.* **2006,** *28,* 1063−1086.

26. Mahaffy, P. G.; Holme, T. A.; Martin-Visscher, L.; Martin, B. E.; Versprille, A.; Kirchhoff, M.; McKenzie, L.; Towns, M. Beyond 'Inert' Ideas to Teaching General Chemistry from Rich Contexts: Visualizing the Chemistry of Climate Change. *J. Chem. Educ.* **2017,** *94,* 1027−1035.

27. Clift, R.; Sim, S.; King, H.; Chenoweth, J. L.; Christie, I.; Clavreul, J.; Mueller, C.; Posthuma, L.; Boulay, A.-M.; Chaplin-Kramer, R.; Chatterton, J.; DeClerck, F.; Druckman, A.; France, C.; Franco, A.; Gerten, D.; Goedkoop, M.; Hauschild, M. Z.; Huijbregts, M. A. J.; Koellner, T.; Lambin, E. F.; Lee, J.; Mair, S.; Marshall, S.; McLachlan, M. S.; Canals, L. M.; Mitchell, C.; Price, E.; Rockström, J.; Suckling, J.; Murphy, R. The Challenges of Applying Planetary Boundaries as a Basis for Strategic Decision-making in Companies With Global Supply Chains. *Sustainability* **2017,** *9,* 279.

28. Towns, M. H. Developing Learning Objectives and Assessment Plans at a Variety of Institutions: Examples and Case Studies. *J. Chem. Educ.* **2010,** *87,* 91−96.

29. Toledo, S.; Dubas, J. M. A Learner-centered Grading Method Focused on Reaching Proficiency With Course Learning Outcomes. *J. Chem. Educ.* **2017,** *94,* 1043−1050.

30. Ambrose, S. A.; Bridges, M. W.; DiPietro, M.; Lovett, M. C.; Norman, M. K. *How Learning Works: Seven Research-Based Principles for Smart Teaching;* Jossey-Bass: San Francisco, 2010.

31. Holme, T.; Murphy, K. The ACS Exams Institute Undergraduate Chemistry Anchoring Concepts Content Map I: General Chemistry. *J. Chem. Educ.* **2012,** *89,* 721−723.

32. Holme, T.; Luxford, C.; Murphy, K. Updating the General Chemistry Anchoring Concept Content Map. *J. Chem. Educ.* **2015,** *92,* 1115−1116.

33. Murphy, K.; Holme, T.; Zenisky, A.; Caruthers, H.; Knaus, K. Building the ACS Exams Anchoring Concept Content Map for Undergraduate Chemistry. *J. Chem. Educ.* **2012,** *89,* 715−720.

34. Luxford, C. L.; Holme, T. A. What do Conceptual Holes in Testing Say About the Topics we Teach in General Chemistry? *J. Chem. Educ.* **2015,** *92,* 993−1002.

35. Meija, J.; Coplen, T. B.; Berglund, M.; Brand, W. A.; De Bievre, P.; Groning, M.; Holden, N. E.; Irrgeher, J.; Loss, R. D.; Walczyk, T.; Prohaska, T. Atomic Weights of the Elements 2013 (IUPAC Technical Report). *Pure Appl. Chem.* **2016,** *88,* 265−291.

36. Harrison, A. G.; Treagust, D. F. The Particulate Nature of Matter: Challenges in Understanding the Submicroscopic World. In *Chemical Education: Towards Research-based*

Practice; Gilbert, J. K., De Jong, O., Justi, R., Treagust, D. F., Van Driel, J. H., Eds.; *Science & Technology Education Library*; Springer: Dordrecht, 2002, Vol. 17.

37. Holme, T. A. Systems Thinking as a Vehicle to Introduce Additional Computational Thinking Skills in General Chemistry. In *It's Just Math: Research on Students' Understanding of Chemistry and Mathematics;* Towns, M. H., Ed.; American Chemical Society: Washington, DC, 2019.

38. Blake, B. Solubility Rules: Three Suggestions for Improved Understanding. *J. Chem. Educ.* **2003,** *80,* 1348–1350.

39. Cannon, A. S.; Finster, D.; Raynie, D.; Warner, J. C. Models for Integrating Toxicology Concepts Into Chemistry Courses and Programs. *Green Chem. Lett. Rev.* **2017,** *10,* 436–443.

40. Masson-Delmotte, V.; Dreyfus, G.; Braconnot, P.; Johnsen, S.; Jouzel, J.; Kageyama, M.; Landais, A.; Loutre, M.-F.; Nouet, J.; Parrenin, F.; Raynaud, D.; Stenni, B.; Tuenter, E. Past Temperature Reconstructions From Deep Ice Cores: Relevance for Future Climate Change. *Clim. Past Discuss. Europ. Geosci. Union* **2006,** *2,* 399–448.

41. Versprille, A.; Zabih, A.; Holme, T. A.; McKenzie, L.; Mahaffy, P.; Martin, B.; Towns, M. Assessing Student Knowledge of Chemistry and Climate Science Concepts Associated With Climate Change: Resources to Inform Teaching and Learning. *J. Chem. Educ.* **2017,** *94,* 407–417.

42. Varma, S. *Preliminary Item Statistics Using Point-Biserial Correlation and P-Values;* Educational Data Systems, Inc.: Morgan Hill, CA, 2006. https://eddata.com/wp-content/uploads/2015/11/EDS_Point_Biserial.pdf.

Using green chemistry to introduce research: Two mini-research projects for the organic laboratory

3

Penny S. Workman, PhD

Associate Professor, Department of Chemistry, University of Wisconsin — Stevens Point at Wausau,
Wausau, WI, United States

3.1 Introduction

Participation in undergraduate research projects is understood to provide numerous benefits to students taking chemistry courses. The American Chemical Society's (ACS) guidelines for bachelor's degree programs recommends undergraduate research for its ability to help students integrate the information they have learned from their coursework and develop their scientific and professional skills *(1)*. The ACS puts such a high value on undergraduate research that it allows up to 180 of 400 required laboratory hours (45%) for ACS certification of a bachelor's degree to come from undergraduate research. Studies on the benefits of participating in undergraduate research include outcomes such as improved scientific literacy, increased engagement with the course and its material, and gains in higher order analytical and critical thinking skills *(2—4)*. Some studies have further suggested that students' greater investment and sense of ownership in their projects increases their interest and persistence in completing majors in chemistry or other STEM disciplines and pursuing science-related careers *(2,5—8)*. In 2012, the President's Council of Advisors on Science and Technology (PCAST) issued the *Engage to Excel* report *(9)*. This report recognized the low degree completion rate of approximately 35% across all STEM disciplines (41.3% in physical sciences) and advocated for a change in the format of undergraduate teaching laboratories, specifically switching from traditional expository "cookbook style" laboratories to true experiments that are based on discovery and research *(2,9)*.

3.1.1 Conventional undergraduate research opportunities

One of the traditional ways in which undergraduate students have participated in research experiences is by joining active research laboratories at their colleges and universities. Summer programs in which students join research laboratories,

Integrating Green and Sustainable Chemistry Principles into Education. https://doi.org/10.1016/B978-0-12-817418-0.00003-6
Copyright © 2019 Elsevier Inc. All rights reserved.

such as Research Experiences for Undergraduates (REU) and Summer Undergraduate Research Fellowship (SURF) programs have also been used to introduce talented undergraduate students to research. While these opportunities are greatly beneficial for the students involved, there are a number of drawbacks to relying on this method as the sole means of providing a research experience to undergraduates. In addition to limited space for undergraduate students in research laboratories and increasing demands on students' schedules that make it harder for them to take advantage of these opportunities, it is often true that students are only recruited to join research groups in the later years of their undergraduate education, typically as juniors or seniors *(2)*. If it is true that participation in research increases the likelihood that an undergraduate student will choose a STEM pathway, waiting until the latter half of their college career to engage them in these opportunities forfeits this benefit. At this point in their academic career, they have likely already selected a major and begun working on the coursework specific to that discipline. Assuming that they participate in a research opportunity at all, the additional time required means that many would still consider it too late to make a change. Indeed, the 2015 publication *Integrating Discovery Based Research into the Undergraduate Curriculum (6)* by the National Academy of Sciences recognized that participation in research opportunities increases retention of students within STEM disciplines and recommended that students be part of these research activities as early as possible in their college careers. Another barrier in this model is the limited or complete lack of availability of research opportunities at small colleges, community colleges and other 2-year institutions *(7)*. Although there has been progress on this front in recent years as 2-year schools have sought to make these opportunities more readily available *(2,5,7)*, the ability of a student to participate in a dedicated research group in the early years of their academic career is still an exception rather than a standard opportunity. While the benefits of exposing students in STEM courses to research early in their academic careers are compelling, at present, these programs are not as ubiquitous and accessible as one might wish.

3.1.2 Approaches to incorporate research into teaching laboratories

An alternate pathway for exposing early undergraduate students to research is to incorporate these experiences into their regular undergraduate chemistry laboratory courses. This approach has the benefit of reaching a much wider number of students with a broader range of interests, including those that may not be currently considering a chemistry major. Methods of incorporating research-like experiences into regular undergraduate courses vary in focus and range from discovery and problem-based experiments in which students explore variables but ultimately derive a known outcome *(10,11)*, and inquiry assignments in which students gather and analyze data in order to answer a specific question *(12)*, to design-based assignments in which students are asked to find or develop their own procedures *(13–15)*, and experiments in which students are actually generating and testing data that could lead to publication. This latter approach is often done in conjunction with an established research laboratory *(2,4,5,16—18)*. While problem-based and inquiry laboratories are not usually considered *authentic* research, in that they do not typically generate new knowledge that is likely to be of interest to stakeholders outside of the classroom *(2)*, they do serve to introduce the students to the idea that work done in a chemistry laboratory is truly an experiment. As the students (and sometimes even the instructors) do not know the outcome of their work ahead of time, this is a real discovery process and requires the students to do a greater degree of critical analysis to reach their conclusions than is typically provided by "cookbook" style chemistry experiments. This serves as a good first step to introducing students to the process of conducting chemistry research and has produced measurable benefits to students. Recent studies have shown these types of laboratory approaches lead to benefits such as increased pass and retention rates, improvements in problem-solving skills and abilities, and gains in students' self-efficacy beliefs and ability to explain experimental results *(11,19—21)*.

3.2 Incorporating green chemistry in the undergraduate laboratory

Over the last 25—30 years, there has been increasing interest in incorporating the ideas of green chemistry into the undergraduate curriculum. Many greener alternatives to traditional laboratory experiments have been developed and made available, and there are now three published laboratory manuals available that focus on green chemistry *(22—24)*. Those working to "green" their laboratories often work to replace existing experiments, especially those that use hazardous chemicals or generate a large deal of waste, with greener alternatives *(25)*. While improving upon the experiments being conducted in the laboratory is certainly an important goal, it is crucial to realize that this is only a first step in incorporating the ideals of green chemistry into the classroom. The need to develop and improve upon chemical reactions and processes to make them less hazardous, more sustainable, and more energy efficient will continue to be an important aspect of doing chemistry in the future. Thus, it is critical to ensure that students are engaged in thinking about and applying green chemistry concepts *(25)*. When properly incorporated into a laboratory curriculum, students will be mindful of the 12 Principles of Green Chemistry *(26)* as they carry out the experiments. They should also be able to analyze a reaction to decide if it is green or if there are serious hazards or nongreen steps or components that could be improved upon.

3.2.1 Bringing students into the process: Prompting analyses and suggestions for improvement

Once students have learned to analyze a reaction to evaluate its greenness, a natural follow-up step is to ask them to suggest specific ways in which a reaction could be improved. Inclusion of this step in a student laboratory setting was described by Bastin and Gerhart *(25)*. Their students were required to include suggestions for ways to improve the greenness of an experiment as part of their laboratory reports. The suggestions were later tested by undergraduate researchers, and if they did prove to be greener, were incorporated into the experimental design in future years. Ribeiro and coworkers have developed the Green Star approach to aid students in analyzing reactions and developing greener protocols to carry them out *(27)*. This method asks students to examine several different procedures for a reaction with respect to each of the 12 Principles of Green Chemistry and individually analyze the greenness of the reaction, isolation, and purification steps from each procedure. The data from all analyses are combined, allowing students to construct and propose an optimized overall procedure. A related project is described by Jessop and coworkers as part of an upper-level course *(28)*. This exercise requires teams of students to analyze four or five methods for carrying out a chemical transformation using nine life-cycle assessment metrics focused on environmental impact. After identifying the greenest method, students use their analyses to propose adaptations to further improve the reaction. Similarly, an online course on green chemistry and sustainability developed by Waddell and coworkers teaches students to use the 12 Principles and green chemistry metrics to analyze a reaction and suggest changes to make the reaction greener. The course culminates with a final project in which students are expected to propose a greener version of an experiment *(29)*.

3.2.2 Green chemistry as a platform for research-based projects

Green chemistry has been identified as a convenient entry point into undergraduate research. An article by Bennett published by the *Council on Undergraduate Research* in 2008 identified a number of benefits in using green chemistry to develop an undergraduate research program including the relatively low start-up costs and reduced safety risks for student researchers *(30)*. Bennett further posited that this type of work would be appealing to students who are interested in participating in work that they perceive as having concrete benefits.

There has been some interest in taking advantage of the opportunities for continued improvement in green chemistry to bring a research element into the regular undergraduate laboratory. This has resulted in the development of protocols for research-based experiments and modules that can be incorporated into existing laboratory programs. These can be divided into two categories. The first category includes inquiry laboratories in which students conduct research-based experiments within the context of green chemistry. Examples of this include Douskey and coworkers' investigative experiment testing antibacterial properties from thyme leaves in general chemistry *(31)*. For this project, students extract essential oils from thyme, form hypotheses related to the oils' antibacterial properties, then test the hypothesis using *Escherichia coli*. Dintzer and coworkers have developed a project in which students attempt the synthesis of substituted tetrahydropyrans using a variety of aldehydes and ketones in a multicomponent reaction using Montmorillonite K10 clay *(32)*. The students first learn the reaction and techniques by completing an expository version of the laboratory. They are then provided with a different aldehyde or ketone and required to research methods to purify their starting materials, then run the reaction following the original experimental design. Once this step is complete, the students work to find ways to optimize the reaction. Mak and coworkers have incorporated green chemistry in an advanced laboratory course designed to introduce undergraduate students to a research environment through the use of room temperature ionic liquids with the Mannich reaction to synthesize piperidone derivatives *(33)*. These students recover the ionic liquids and repeat the procedure twice, comparing the results of each reaction. The results are also compared with those obtained using a traditional organic solvent. An investigation using fluorous solid phase extraction (FPSE), a greener method of performing solid phase separations, has been developed by Slade, Pohl, and coworkers *(34)*. Students perform literature searches to find procedures for each step in a multistep synthesis of a fluorous dye molecule and adapt them for use in their laboratory. The products are purified using FPSE. Gross et al. have developed an investigative laboratory in which students research procedures for biodiesel synthesis, both from websites and from the chemical literature, and use them to develop their own procedures to test variables in the synthesis *(35)*. The students perform numerous tests to analyze the resulting biodiesels. Lee et al. have described a multisemester project incorporating faculty research interests in asymmetric reduction of aldehydes and ketones *(36)*. In the first semester, organic chemistry students were challenged to develop a synthesis for a ketone that was not commercially available. In the second semester, the students are provided with unknown ketones and use polylactic acid derivatives as chiral aids in reducing the ketones. The products were analyzed to determine whether chiral reduction had occurred.

The second category includes laboratory investigations in which greening the experiment is the research focus. This category includes Dicks and coworkers' experiment for upper-level organic chemistry students in which the students develop their own synthesis of an azlactone using green chemistry in their designs *(37)*. Similarly, Bastin and Gerhart have designed a first-semester capstone project consisting of a multistep green synthesis. Students in this course are expected to use the literature to locate procedures for each step of the transformation. The class analyzes each procedure for cost-effectiveness, safety, and greenness and selects the best procedure for each step. If the analysis suggests that a proposed procedure that has not been tested may be greener, the student attempts the synthesis using that method *(25)*. Graham and coworkers have developed a student-designed green chemistry research project in which organic chemistry students choose a synthesis from the chemical literature and develop an alternate procedure meant to make the reaction greener *(38)*. Students work in pairs, with each student performing the reaction once as written and once with the green modifications.

3.3 Overview of green chemistry mini-research projects for the organic laboratory

Two mini-research projects focused on green chemistry in organic laboratories have been developed at the University of Wisconsin–Stevens Point at Wausau. The first project challenges students to develop greener versions of organic reactions commonly found in a published laboratory manual. The second project focuses on the use of renewable and recycled chemical feedstocks. Both projects were initially developed as a laboratory component for an honors organic chemistry course. To date, one of these projects has been adapted for use in the regular organic chemistry laboratory. Both projects are described in this chapter, along with the steps and assignments developed to prepare students to undertake the projects.

The University of Wisconsin–Stevens Point at Wausau is a 2-year campus in central Wisconsin that is part of the University of Wisconsin System. It has an annual enrollment of approximately 700 students. The campus grants associate degrees; however, its primary mission is to provide the first two years of the under-graduate curriculum and prepare students to transfer to a 4-year institution to complete their bachelor's degree. The organic chemistry laboratory is a stand-alone, two-semester course that meets for one 2-h and 50-min session each week. Students enrolled in the laboratory must either be concurrently enrolled in the organic chemistry lecture or have completed it in a previous semester. In an average year, approximately 15 students enroll in the organic chemistry laboratory. The honors organic chemistry course is a separate one-credit course that is offered in the Spring semester. Students are encouraged to enroll in the honors course if they have earned an A or B grade in the first semester of organic chemistry lecture and laboratory and are either concurrently enrolled in the second semester of both courses or have completed them in a prior semester. The enrollment in the honors organic chemistry course has varied from one to seven students, although two to four students is typical.

3.3.1 Project goals and objectives

Both projects share an overarching goal of familiarizing students with green chem-istry and its practice and providing an introductory experience to how chemistry research is conducted. In order to achieve this goal, students must successfully meet a number of skill objectives. These objectives are shown in Fig. 3.1.

3.4 Preparative assignments

Preparative assignments have been developed for use throughout the year to assist students in developing skill objectives 1–4. Analysis and interpretation of experi-mental data (skill objective 5) are incorporated into each laboratory exercise and its post-laboratory assignment. At present, the only student presentation in the

---

Skill Objectives for Green Chemistry Mini-Research Projects

1. Access and use Safety Data Sheets and other safety information to determine risks and handling precautions for chemicals used in an experiment.

2. Use the 12 Principles of Green Chemistry and green chemistry metrics to analyze reactions for greenness.

3. Adapt information and procedures to design and carry out an experiment.

4. Locate, access, read, and interpret journal articles.

5. Analyze data to determine the success of their experiments.

6. Present the results of laboratory work to classmates and chemistry staff.

FIGURE 3.1

Skill objectives for mini-research projects.

course is the project presentation; however, additional presentations could be added throughout the semester to give students an opportunity to develop these skills.

Table 3.1 lists the preparative assignments, their associated skill objectives, and the approximate timeline for when the assignments are given. Each of the assignments is discussed in the following paragraphs.

Table 3.1 Correlation of preparative assignments to objectives and placement in course.

Assignment	Skill objectives	Placement in course
Chemical safety assignment	1	Fall semester, week 1
Comparison of traditional and green extraction methods	2	Fall semester, week 6
Synthesis of butyl naphthyl ether/introduction to green chemistry metrics	2	Fall semester, week 9
Adapting dehydration procedure	3,4	Fall semester, weeks 9–10
Reaction design project	3	Fall semester, weeks 10–15
Chemical literature assignment	3, 4	Spring semester, week 3
Pre-laboratory assignments	1, 2	Throughout course
Post-laboratory assignments	2, 5	Throughout course

3.4.1 **Chemical safety**

The first step in preparing students for these projects is to help students learn about chemical safety. Arguably, this is a critical goal for any chemistry laboratory. It is, in fact, one of the learning objectives for college students identified by the ACS *(39)*, and various approaches to achieving this goal have been described in the literature *(40–48)*.

Due to its critical nature, this assignment is part of the first laboratory session. The students are divided into groups of three. Each group is assigned three different chemicals that they must investigate: one that is fairly benign, one that presents a moderate hazard, and one that is a more serious hazard. The students are provided with background information about Safety Data Sheets (SDSs), the National Fire Protection Association Hazard Identification System (NFPA fire diamond), and Hazard Communication (HazCom) pictograms. The students must fill out a worksheet for each of their assigned chemicals that asks about information that can be obtained from each of these resources. After researching each of the assigned substances, the students fill out a summary sheet that asks them to compare the three chemicals. The worksheet and summary sheet are available in Section 3.8.1.

Students are directed to use their phones, laptops, or university computers to complete the assignment during the laboratory period. The assignment takes about an hour. Once the groups have finished, the instructor holds a full-class discussion to review the information and how the resources can be used to evaluate a chemical's safety and handling procedures. The discussion makes it easier to identify any deficits in the students' understanding and allows the instructor to quickly address any problems or misconceptions.

After this introductory assignment, questions about chemical safety become a regular part of pre-laboratory assignments. Students are frequently required to find and evaluate SDS and HazCom data for the materials they will be using. Incorporating these searches into the pre-laboratory assignments reinforces the material and emphasizes the importance of knowing this information before handling a chemical.

3.4.2 **Introduction to green chemistry**

Students are first introduced to green chemistry about one-third of the way through the Fall semester. This is incorporated as part of the pre-laboratory reading for an experiment comparing two different methods of extracting the essential oil (*R*)-(+)-limonene from orange peels: a traditional steam distillation-extraction such as the one found in the text by Mohrig *(49)* and a liquid carbon dioxide extraction *(50)*. The reading provides a brief overview of green chemistry and a list and short explanation of each of the 12 Principles of Green Chemistry. Students perform both procedures. The post-laboratory questions ask them to use the 12 Principles to identify ways in which the liquid carbon dioxide extraction is greener. The hands-on experience with both procedures makes it relatively easy for students to answer this question. Typical answers include less water waste, less energy consumption, and elimination of the hazardous organic solvent and other materials needed in the extraction and drying steps of the more traditional procedure.

The first synthetic reaction, preparation of the artificial flavoring butyl naphthyl ether via an S_N2 mechanism ((*51*), Scheme 3.1)), is performed around week nine of the fall semester. This reaction uses ethanol as the reaction solvent instead of the less benign polar aprotic solvents the students have studied in lecture. As part of their pre-laboratory assignment, students explain how this procedure relates to Principle 4: designing benign chemicals and Principle 5: use of benign solvents and auxiliaries. The use of green chemistry metrics is introduced in the post-laboratory assignment. Students are asked to calculate the atom economy and experimental atom economy for the reaction, and explain why the two calculations give different results. Two more metrics, effective mass yield and E factor, are introduced later in the semester.

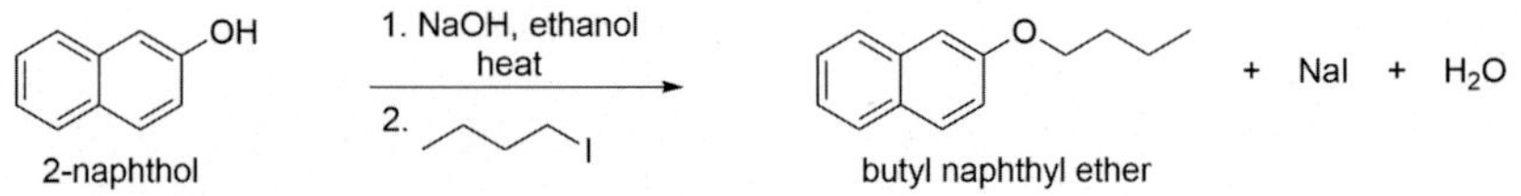

2-naphthol

1. NaOH, ethanol
   heat

2.

butyl naphthyl ether   +   NaI   + H_2O

SCHEME 3.1

S_N2 reaction to produce butyl naphthyl ether.

Once green chemistry has been introduced, questions requiring students to analyze the laboratory experiments using the 12 Principles of Green Chemistry or one or more green chemistry metrics become a regular part of pre- and post-laboratory assignments. This often requires them to compare the reaction they performed to a more traditional, less green version. These comparisons help them gain a broader perspective of the many ways a reaction can be improved upon. By the end of the fall semester, students are expected to be able to suggest specific changes that would improve upon the greenness of the reaction performed as part of their post-laboratory assignments.

3.4.3 Adapting procedures

The third skill objective (learning to adapt a published procedure) is developed through three different assignments throughout the course of the academic year. The first assignment, adapting a dehydration procedure, is introduced during week nine of the fall semester. This is followed by the end-of-semester project in which students locate and adapt a procedure for a selected starting material. This skill is revisited again in the first few weeks of the spring semester as part of the chemical literature activity.

3.4.3.1 Adapting a literature procedure Assignment 1: Dehydration of cyclohexanol

In the first of the adaptation assignments, the class is provided with a very short journal article describing the dehydration of 2-methylcyclohexanol ((*52*), Scheme 3.2). The students are told that in the next laboratory period, they will be dehydrating

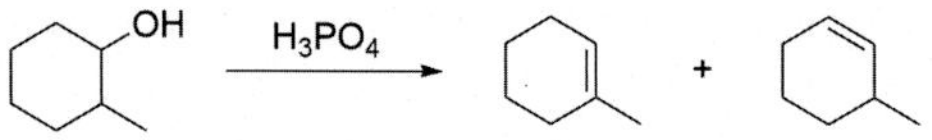

SCHEME 3.2

A published procedure for the dehydration of 2-methylcyclohexanol.

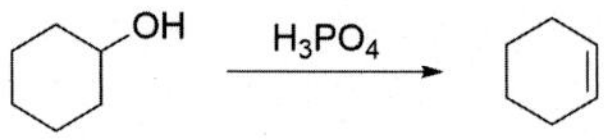

SCHEME 3.3

A modified reaction for the dehydration of cyclohexanol.

5 mL of a chemically similar molecule, cyclohexanol (Scheme 3.3). They are responsible for using the article provided to write a procedure for the experiment, which must be turned in prior to beginning laboratory work. The original experiment uses not only a different starting material but also a much larger scale: 20 g of 2-methylcyclohexanol. In order to successfully prepare their procedures, the students need to determine what adjustments must be made in order to account for the differences between the two starting materials. Specifically, they must consider both molecular mass and density of the starting materials in order to scale the reaction and determine the amounts of other reagents to use and to make good choices regarding glassware sizes. The purification step in this reaction is a distillation, meaning that it is also necessary to consider what changes need to be made as a result of the difference in the boiling points of the final products. The instructor provides guidance on making the necessary changes as part of a class discussion when the assignment is first made available. Before beginning their procedures the next week, students are assigned to groups to discuss and compare their procedures with their classmates and resolve any differences. This discussion usually suffices to find and correct any individual mistakes. The instructor circulates through the laboratory and discusses the procedures with the groups before they begin, intervening and providing further assistance when necessary.

3.4.3.2 End of semester project: Locating and adapting a published procedure

The end of semester project is an open-ended experiment that allows students to apply and reinforce the skills they have learned about adapting procedures. The assignment details are provided immediately after the dehydration experiment is completed. Students work individually and are provided with a choice of four possible starting materials. They are tasked with locating and adapting a procedure to use their chosen starting material in a synthetic reaction. The criteria used by the instructor for choosing potential starting materials are shown in Fig. 3.2.

The first two criteria are intended to make the task more manageable. Functional groups are limited to those that have been covered in the first semester organic chemistry lecture course. This means that students already have some familiarity with the

---

Criteria for Starting Materials

1. Contain functional groups that are familiar to students.

2. Contain only one reactive functional group.

3. Not commonly used in published laboratory manuals.

4. Easy to handle.

FIGURE 3.2

Instructor criteria for selection of starting materials.

---

functional groups and their reactivities. While students are not limited to the reactions studied in lecture, this often provides a good starting point. Starting materials contain only one reactive functional group in order to avoid undesirable side reactions, many of which students would not yet have the background knowledge to predict. This assignment allows the use of procedures from published laboratory manuals, so it is important to avoid starting materials that are commonly used in the manuals. Finally, the starting materials must not require any special handling techniques unfamiliar to students. Alkyl halides, alkenes, and alcohols are frequently used as starting materials for this project.

The students are allowed up to 5 mL of their chosen starting material. A published laboratory manual or journal article must be used as the source for their procedures. Procedures that have been used in the course earlier in the semester are not allowed. Most students opt to use laboratory manuals because they do not yet have much experience searching for and using journal articles. The students have two weeks to find a reaction and submit a plan that includes a detailed procedure for carrying out the reaction, including isolation and analysis of the product. They must also submit complete reference information for all sources. This is good practice and also allows the instructor to easily access the source if the proposal appears to have errors. The assignment sheet given to students is available in Section 3.8.2.

When explaining the assignment, the instructor advises the students that they are unlikely to find a procedure using the exact starting material and discusses the considerations that will be helpful in finding a useful procedure. If the starting material is an alcohol, for example, it is useful to consider the type of alcohol (primary, secondary, or tertiary) and search for a procedure that also uses an alcohol of this type. Students are also reminded to pay attention to the physical properties of their starting materials and products so that they can propose appropriate isolation and analytical procedures. This is important as the most common error in the proposals typically stems from proposing a workup or analysis inappropriate to the physical state of the starting material, e.g., proposing distillation of a solid or obtaining the melting point of a liquid.

Student proposals are reviewed by both the instructor and the laboratory manager. If corrections are necessary, the instructor provides assistance in working through any problems. If a student has proposed a reaction that cannot be carried out due to cost, safety, or the inability to obtain chemicals or equipment, the instructor works with the student to find another suitable procedure. As the submitted procedure fulfilled the requirements of the design portion of the assignment, the student is awarded credit for this portion even though they will actually perform a different procedure.

Students have the final two weeks of the semester to carry out their reactions. Most students complete their experimental procedures in the first week and only need to do analysis during the second week. In addition to their laboratory reports, students must turn in the answers to two post-laboratory questions. The first question asks students to reflect on their reaction, procedure, and results and make suggestions on what they would change in order to improve the reaction if they were going to repeat the procedure. This is intended to introduce the idea that a single iteration of a reaction is rarely a reality outside of a teaching laboratory. In both research and industrial settings, reactions are typically repeated to optimize the reaction. From the student perspective, this is a logical next step following discussion of experimental errors, and they are usually able to identify several specific changes that would improve the reaction. The second question asks students to suggest a change that would make the reaction greener. While green chemistry is not a specific goal of this assignment, this question is included with the hope that continued engagement with green chemistry concepts will help prepare them for their project in the second semester.

3.4.3.3 Accessing and using chemical literature

The final preparative assignment focused on adapting procedures is incorporated into a chemical literature assignment. This is described more fully in Section 3.4.4 and focuses primarily on teaching students to locate and read journal articles. The last portion of this assignment requires students to use the supporting information for a specific article to adapt a procedure for use with a chemically related starting material. They must also explain why it is reasonable to expect that the procedure will work with the new starting material.

3.4.4 Accessing chemical literature

The goal of learning to access the chemical literature is incorporated during the first three weeks of the spring semester through the use of a take-home activity designed to help students learn how to navigate the ACS's publications page and use it to locate and access journal articles. Prior to giving the assignment, the instructor holds a short discussion to introduce students to the website and briefly demonstrates the use of the search function. Students practice reading abstracts to help quickly determine whether an article is of interest and are shown how to access the supporting information for an article. Although this assignment focuses on ACS journals, students are also shown how to use the Web of Science database, how to access Royal Society journals through the library website, and how to request other articles through an interlibrary loan. There are a number of recent examples in the literature of other assignments and strategies developed to teach information literacy as part of a laboratory course *(53−61)*. Many of these assignments include instruction for the use of SciFinder. Unfortunately, SciFinder access is not available at University of Wisconsin−Stevens Point at Wausau; however, the available resources are sufficient to meet the goal of introducing students to the process of locating and accessing articles.

The literature assignment teaches students to use the ACS publication page to locate recent articles in specific journals, conduct searches on topics of interest, use the advanced search feature, and conduct a citation search to find a specific article. The students must read the specified article and its supporting information, then answer specific questions that help them learn how to find and interpret data from reaction schemes and experimental sections. Finally, they must use the reaction details presented in the supporting information to adapt the procedure for use with a related chemical, as described in Section 3.4.3.3. The assignment is shown in Section 3.8.3.

The chemical literature assignment is the last of the preparative assignments; however, students continue to practice analyzing reactions using the 12 Principles of Green Chemistry and green chemistry metrics through pre- and post-laboratory questions throughout the spring semester.

3.5 Description of projects

Both projects were initially developed as part of the honors organic chemistry course. The first project focuses on applying the 12 Principles of Green Chemistry and green chemistry metrics to published laboratory activities in an attempt to improve upon their greenness. This project has been adapted for use in the regular organic chemistry laboratory course. The second project focuses on the use of renewable and recycled chemical feedstocks. Both projects are discussed in this section, as are the adaptions necessary for implementation of the first project in the regular laboratory.

3.5.1 Project 1: Greening a published laboratory experiment: Honors section

Students in the honors section were given the task of selecting an experiment from a published laboratory manual and finding ways to make the reaction greener. The students worked individually, giving each student complete ownership of their project. After selecting an experiment, the students were required to identify two different changes that had the potential to improve the reaction's greenness. Each change was required to correspond to a different green chemistry principle. This requirement was intended to ensure that the two proposals were substantively different, eliminating a problem that emerged during the first iteration of the project in which some students proposed two changes that were essentially the same, e.g., two new reactions in which the only difference was solvent choice. This requirement also forced students to consider multiple aspects of green chemistry. Finally, the students were required to make use of published journal articles to develop procedures for each of the proposed changes, including plans for purification and analysis of their products. The students turned in proposals complete with full procedures and lists of the required chemicals and glassware. The required amount of each chemical needed to be included in the list. The assignment handout is shown in Section 3.8.4.

Proposals were reviewed by both the course instructor and the laboratory manager to verify their safety, feasibility, affordability, and potential to improve the greenness of the reaction. If a proposed experiment included use of substances deemed too hazardous for student use or failed to meet the objective of producing a greener reaction, the student was required to either submit a revision or a new proposal.

After approval of the proposals, students scheduled individual laboratory time to run their reactions. They were required to first run the reaction as written in the laboratory manual, then run each of the proposed reactions. After obtaining the results of their work, students analyzed all three reactions using the 12 Principles of Green Chemistry and four green chemistry metrics: atom economy, experimental atom economy, E factor, and effective mass yield. The students presented their projects to their classmates and members of the Chemistry Department in a 10–15 min PowerPoint presentation, followed by a short time for questions. The requirements for the presentation are included in the assignment handout (Section 3.8.4).

3.5.1.1 Timeline for honors section

The honors organic chemistry course was divided into two parts. For the first nine weeks of the semester, the course met for one 50-min lecture session each week. During the remainder of the semester, students scheduled individual laboratory time to complete the mini-research projects. The assignment details for the project were first provided to the students during the first week of March. This coincided with the end of the green chemistry unit in the lecture portion of the course.

The students had three weeks to develop their proposals. This allowed two weeks for instructor review of the proposals and purchase of materials. Students were able to schedule and perform laboratory work throughout the month of April. The amount of time necessary for each student varied in accordance with their procedures. Most students required either two or three 3-h time blocks. Some students proposed reactions that required longer time blocks, and a few students required a fourth time block. The students presented their work during the last week of classes.

A timeline showing the project in correlation to the course as a whole is shown in Fig. 3.3.

3.5.1.2 Example of an honors section project

One student proposed to examine the photobromination of 1,2-diphenylethane. The original reaction *(49)* is shown in Scheme 3.4.

Unit	Weeks	Topics
Green Chemistry	2-4	Introduction to Green Chemistry
		Applied Green Chemistry / Presidential Green Chemistry Award
		Green Chemistry Metrics
Drug Discovery	5-9	Taxol (How a Tree Becomes a Drug)
		Natural Products: Isolation and Structure
		Determination Advanced Analytical Methods
		Total and Semi-Synthesis
		Drug Development and Testing
Laboratory Work	10-14	
Presentations	15	

FIGURE 3.3

Timeline for the honors course.

$$KBrO_3 + 6HBr \longrightarrow 3H_2O + 3Br_2 + KBr$$

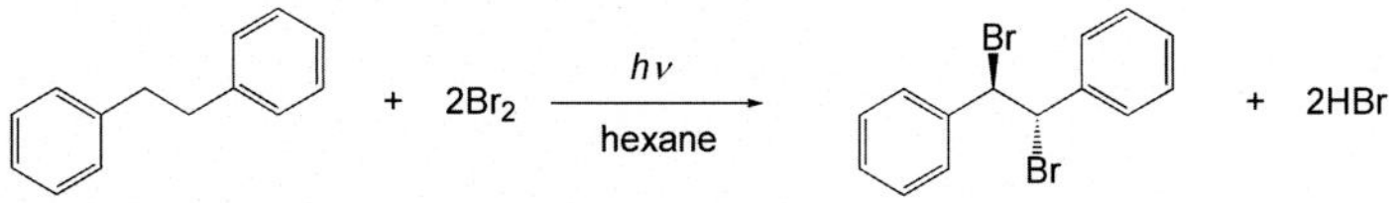

SCHEME 3.4

Unmodified photobromination of 1,2-diphenylethane.

Although this reaction does employ in situ generation of bromine, thus improving immediate safety hazards in comparison to using bromine directly, the text acknowledged persistent safety concerns. Potassium bromate is a suspected carcinogen. The strong oxidizing nature of the compound also required special handling and disposal instructions to avoid a potential risk for fire.

The student's first proposal, shown in Scheme 3.5, offered an alternate way to generate bromine in situ. The student proposed the use of N-bromosuccinimide (NBS) in place of the potassium bromate from the original reaction (62). NBS is corrosive, but is not suspected of carcinogenicity.

This reaction produced a lower yield, 17% as compared to the 42% obtained by the student from the original reaction. The products gave similar IR spectra; however, the melting point of the product from the modified reaction was lower than that of the original, 200°C (dec) as compared to 242−243°C (dec). The student was unable

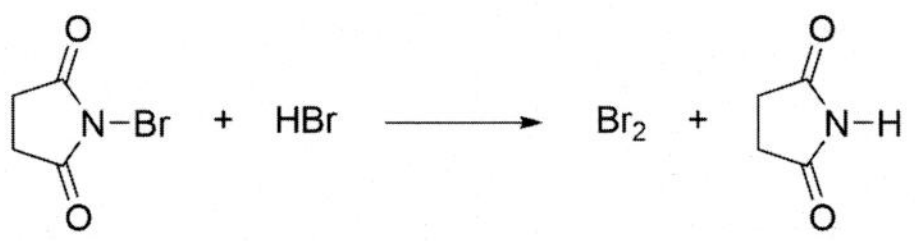

SCHEME 3.5

Proposal 1: *In situ* generation of bromine using *N*-bromosuccinimide.

to state definitively whether or not the reaction had been successful, but concluded that it "probably did not work" based on the melting point data. The student did, however, note that further experimentation could yield better results.

The student's second proposal focused on changing the reaction solvent. The student proposed substituting hexane from the original experiment with 2,5-dimethyl-hexane *(63)*. While this solvent is still not very green, the student was able to show some small improvements with respect to health hazards and was allowed to make the substitution. The percent yield in this reaction was 62%, higher than that of the original reaction. The IR spectrum was similar to the original product, and the melting point was 240–241°C (dec). The student was able to conclude that the reaction was successful and had improved upon the yield of the original reaction, although further experimentation would be needed to validate the improved outcome.

3.5.2 Project 1: Greening a published laboratory experiment: Regular organic laboratory

Due to the larger class size, several modifications were necessary in order to incorporate this project into the regular organic chemistry laboratory. These changes were largely made to make the projects more manageable and to limit the costs associated with running open-ended experiments. Instead of working individually as the honors students had done, students were assigned to work with a partner. The instructor selected a single experiment that was used by all groups. Each pair was required to propose one change to improve upon the greenness of the reaction. The change to just one proposal from each pair was made not only to limit costs but also to ensure that all experimental work could be completed within two laboratory sessions.

Like the honors students, students in the regular laboratory section were required to use journal articles to develop their procedures. The proposal requirements were unchanged and can be viewed in the description of the honor assignment in Section 3.8.4. The students presented their work in 10–15 min PowerPoint presentations using the same requirements used in the honors course.

3.5.2.1 Timeline for the regular course

Project details were provided to the students around the second week of March. Proposals were due at the end of the first week of April, which corresponded to week 9 of the 15-week semester. The proposals were reviewed by the instructor and laboratory manager the following week. This allowed two additional weeks for any necessary revisions to the proposals and purchase of required materials before experimental work began. The last three weeks of the semester were dedicated to the project. All experimental work was conducted during weeks 13 and 14. Presentations were given during the final day of class in week 15.

Fig. 3.4 shows a typical laboratory schedule for both fall and spring semesters. Experiments with green chemistry components are marked with an asterisk. Experiments that contribute to the other goals of preparing students for a mini-research project are marked with a double asterisk. Project information is in italics.

	Fall Semester	Spring Semester
Week	Experiment	Experiment
1	Lab Check-In / Safety Assignment	Grignard Synthesis of Crystal Violet and Malachite Green
2	Melting Points	Organozinc Reaction*
3	Recrystallization	Introduction to Chemical Literature
4	Boiling Points and Refractive Index	Stereoselectivity of Borohydride Reduction of Benzoin
5	Thin Layer Chromatography	Identification of an Unknown
6	Isolation of (R)-(+)- Limonene from Orange Peels*	Synthesis and Chemiluminescent Reaction of Divanillyl Oxalate *Project Assignment Handed Out*
7	Physical Properties of Enantiomers	Diels-Alder Reaction
8	Nucleophilic Substitution	Multi-Step Synthesis Week 1
9	Synthesis of Butyl Naphthyl Ether*	Multi-Step Synthesis Week 2 *Project Proposal Due*
10	Elimination**	Wittig Reaction*
11	Acid-Base Extraction	Synthesis of Adipic Acid*
12	Stereochemistry of Bromine Addition to *trans*-Stilbene*	Saponification
13	Isolation and UV-Visible Spectroscopy of Lycopene from Tomato Paste	*Green Chemistry Project Lab Work*
14	Procedural Design Project**	*Green Chemistry Project Lab Work and Analysis*
15	Procedural Design Project**	*Presentation of Projects*

FIGURE 3.4

Organic chemistry laboratory schedule with project information.

3.5.2.2 Example of a project in the regular laboratory section

The assigned experiment was the photobromination of 1,2-diphenylethane laboratory described in Section 3.5.1.2. The reaction is shown in Scheme 3.4. Proposals from the regular laboratory have tended to avoid making changes to reagents and are more likely to focus on solvent choice or energy consumption. The original reaction *(49)* made continuous use of a 100 W incandescent bulb to form bromine radicals. One group proposed using sunlight as a natural light source *(64)* and set up their reaction near the laboratory window. Both reactions were allowed to stir near their respective light sources until the orange color of the bromine had dissipated and a white precipitate had formed. The group obtained an 84% yield from the original reaction with a melting point of 236°C (dec). The yield for the modified reaction was 18%. The melting point was 212°C (dec). The group speculated that the lower melting point could be the result of impurities in their product.

Another group chose to substitute heptane for the original hexane solvent. This swap is consistent with the solvent classifications in the GlaxoSmithKline (GSK) solvent selection guide *(65)*. This group reported a 77% yield and a melting point of 229–237°C (dec) for the original reaction and a yield of 60% and melting point of 231–238°C (dec) for the modified reaction.

3.5.2.3 Reflections on Project 1

As indicated in the project description, adapting this project to the regular organic chemistry laboratory required some adjustments and modifications to allow the project to be feasible with the larger class size and more restrictive course schedule. It was necessary to have students work in pairs and to restrict the experiment options for the larger class to make the project more manageable and economical. The scheduling of the regular organic laboratory section further restricted options for this project. The honors students had a large degree of flexibility in scheduling. Their ability to schedule laboratory time was limited only by their schedules and the availability of the instructor or laboratory manager, who helped supervise the projects. This allowed for reaction modifications that required longer reaction times and steps that needed to be completed over a series of several days. In the nonhonors course, students were restricted by the laboratory schedule and needed to complete their work within two laboratory periods, each lasting 2 h and 50 min. Exceptions were made for reactions that could be set up in advance and allowed to run safely without constant supervision. This was possible as there is only one section of the organic chemistry laboratory and would likely be more difficult at a larger school with multiple laboratory sections sharing space and equipment.

While students in the honors course could choose from any of the synthetic reactions in their laboratory text, the most popular choices were always bromination reactions. This was likely due to the inclusion of the Hutchinson article on greener bromination of *trans*-stilbene in the lecture portion of the course *(66)*. While they were not permitted to use the approaches discussed in the article, the students recognized brominations as reactions with significant room for improvement and were able to find other ways to modify the reactions to make them greener. The most frequently proposed changes included finding other methods to generate bromine in situ and experimenting with greener solvents. In general, most students ended up with one reaction that successfully produced the desired product and one that did not. Due to its popularity and the relative success that was achieved by the honors students, the photobromination reaction was selected for use in the regular laboratory section. The only green modification that students were not allowed to propose was generation of bromine in situ using hydrogen peroxide and hydrobromic acid. This procedure had been used during the first semester of the course in performing the green bromination of *trans*-stilbene *(66)*. Since performing a literature search was one of the goals of the project, this reaction was excluded. As noted earlier, the students in the regular laboratory tended to favor simpler modifications that investigated greener solvents and alternate light sources. This may be a result of self-selection. It may be that students that choose to take the honors course are more likely to be interested in a research-based project and thus more likely to investigate more complex changes.

It is important that students understand that the project grade does not depend on the success of the reaction. Failed reactions are an inevitable part of research and can provide useful information about the reaction. The literature sources used by the students must provide sufficient evidence that the proposed reactions are reasonable. As long as these criteria are met, an unsuccessful reaction is not counted against the project grade. While students are understandably disappointed if their modifications do not work, knowing that the reaction's failure will not negatively impact their grades alleviates some of the stress associated with this project. The biggest impact of a failed reaction is the additional work required as the students prepare the future work portion of their presentation.

3.5.3 Project 2: Chemical Feedstocks

The second project focuses on the use of alternate chemical feedstocks that are obtained either from renewable sources or postconsumer waste materials. The students were tasked with isolating a chemical from one of these sources, then demonstrating its viability as a feedstock by using the chemical in a subsequent reaction. Thus far, this project has only been implemented in the honors organic chemistry course.

3.5.3.1 Project description

Students worked individually to complete this project. After the students selected a renewable resource or postconsumer waste product, their next task was to use the chemical literature to locate a procedure that would allow them to either isolate a compound directly from the source or to perform a chemical reaction that would result in the isolation of a single compound. The second task required students to use journal articles to prepare procedures that would allow them to use their compound in a synthetic reaction, thus demonstrating the viability of their source as a feedstock. As was the case in the original green chemistry project, students were required to prepare and submit proposals that included detailed procedures, lists of all required materials, and full reference information for all sources. The full project description is shown in Section 3.8.5.

The student proposals were reviewed by both the course instructor and the laboratory manager to ensure their safety, feasibility, and affordability. After the projects were approved, students scheduled individual laboratory time to isolate their initial compounds, carry out their syntheses, and perform all analyses. The students presented their work to their classmates and members of the Chemistry Department in a 10−15 min PowerPoint presentation, followed by a short time for questions. The requirements for the presentation are included in the project description.

3.5.3.2 Timeline for chemical feedstocks project

The timeline for the chemical feedstocks project was identical to the timeline for the original project in the honors course (Fig. 3.3). The students had three weeks to develop their proposals. The instructor and laboratory manager had two weeks to review the proposals and obtain the materials needed for the procedures. Students were able to schedule and perform laboratory work throughout the month of April. The amount of time necessary for each student varied in accordance with their procedures. Most students required only two 3- or 4-h time blocks. Some chose to schedule smaller separate time blocks to conduct analyses. The students presented their work on the last day of class.

3.5.3.3 Example project

One student chose to use Styrofoam packing material as a feedstock. This student chose to do a reaction directly with the Styrofoam, isolating a product that would then be used in a subsequent reaction. The student performed the sulfonation reaction shown in Scheme 3.6 (67). The student was interested in performing this reaction because the sulfonated product has been shown to be useful in the removal of some heavy metals from water.

An IR spectrum of the student's product was consistent with the IR data provided by the article, indicating that the reaction was successful. The percent yield for the reaction was 97%. The student's follow-up reaction was a methylation using trimethylorthoformate. (68). The proposed reaction is shown in Scheme 3.7. This reaction was not successful. Interestingly, in spite of the IR data that showed the first reaction had been successful, the student suggested that neither reaction had worked in his presentation. This conclusion was reached based on the failure of the second reaction and the student's mistrust of the dark color of the first product.

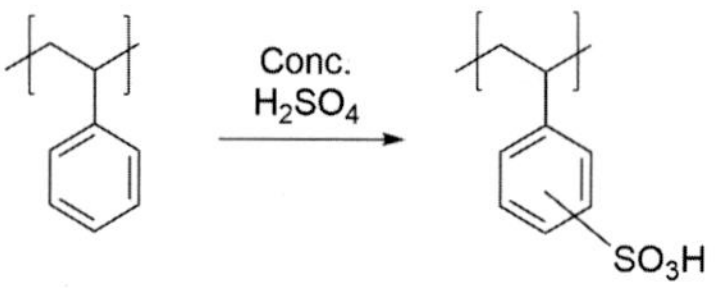

SCHEME 3.6

Sulfonation of Styrofoam.

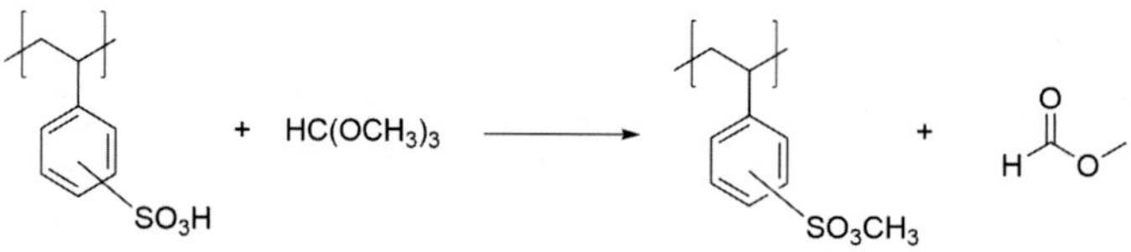

SCHEME 3.7

Proposed methylation reaction.

3.5.3.4 Reflections

The chemical feedstocks project has been enthusiastically embraced by the honors students. When this project was first initiated, it was done with the expectation that students would start in fairly obvious places and perform very simple transformations. For example, an experiment that began with the fermentation of sugar to create ethanol followed by oxidation to produce acetic acid would fulfill the project criteria and be relatively easy for the students to think of and find procedures for. Every student who has completed this project has far exceeded these expectations. The proposals have called for starting materials such as corn, peppermint leaves, Styrofoam, and plastic cups and included procedures meant to synthesize useful products such as insect repellants, biodiesels, compounds used for water purification, and fire-resistant polymers. Others have sought to explore advanced reactions well beyond the scope of most sophomore-level chemistry courses. These students seem eager to be able to do work beyond traditional experiments. Several have commented that the opportunity to participate in the self-directed project was the biggest factor in their decision to take the honors course.

3.6 Future directions

One component that is still missing in these experimental designs is the iterative feedback cycle that occurs in real research. While this would be more difficult to implement in the one-semester honors course, beginning the project earlier in the regular laboratory course could provide students with the opportunity to use their previous results as data and try out some of their ideas to modify and improve upon their experimental reactions and designs. Alaimo, Langenhan, and Suydam described a "divide-and-conquer" approach to implementing data-driven investigations in which several groups of students each focused on a different reaction condition in order to optimize the reaction *(69)*. The data from each group were compiled and shared with the class. Adopting this approach could allow for incorporation of an optimization cycle and a concluding experiment in which students use class data to attempt a reaction that has made improvements upon the greenness of the reaction in multiple areas.

As explained earlier, the project exploring the use of renewable and recycled feedstocks has not yet been adapted for use in the regular laboratory. Future work will include adapting the feedstocks project in a manner that will make it possible to offer both projects as an option in the regular course in hope that allowing students to select the project that they are more interested in will increase their interest and engagement with the project.

3.7 Conclusions

The two projects described in this chapter provide organic chemistry laboratory students with an introduction to conducting chemistry research. The theme of green chemistry provides a convenient entry point that is accessible to undergraduate students at the sophomore level. Preparation for the projects includes numerous assignments over the course of a two-semester laboratory sequence that teach students the skills necessary to carry out independent investigations. The ongoing, consistent focus on the 12 Principles of Green Chemistry and green chemistry metrics throughout the course helps to ensure students are mindful of safety concerns and cognizant of the need to develop more sustainable chemistry.

Both projects, along with the preparatory assignments, are meeting the skill objectives for the projects. By the end of the second semester, the students are able to access safety information and most students have demonstrably improved in their abilities to determine which chemicals present special hazards and how they should be handled. The students are able to locate published journal articles that are relevant to their reactions and adapt the procedures to their specific materials. Other than small procedural changes, they are able to do this on their own without instructor intervention. Finally, they are able to analyze their reactions and provide evidence as to whether their changes were successful in improving the greenness of the reaction and make concrete suggestions on how the procedures could be further improved.

3.8 Student handouts

3.8.1 Chemical safety assignment

Chemical safety

Safety Data Sheets (SDS) are important documents that are required, by law, to be kept on file for each chemical used at a workplace. The SDS contains important information about the chemical, including its physical properties, reactivity, and safety information. Having quick access to this data can provide crucial information on how to store a chemical, clean up a spill, or treat a person exposed to the chemical. It is very important to be familiar with this information when working with a chemical.

The Occupational Safety and Health Administration (OSHA) requires certain information to be present in an SDS for every chemical. Although not strictly required, OSHA recommends that the writers of SDSs follow the format set forward by the American National Standards Institution (ANSI). ANSI's format for SDSs has 16 sections:

- Identification
- Hazard(s) identification
- Composition/information on ingredients
- First-aid measures
- Fire-fighting measures
- Accidental release measures
- Handling and storage
- Exposure controls/personal protection
- Physical and chemical properties
- Stability and reactivity
- Toxicological information

- Ecological Information
- Disposal considerations
- Transport information
- Regulatory information
- Other information

For many years, important chemical safety information was also included on National Fire Protection Associate (NFPA) labels, also known as the fire diamond. While this system has recently been changed to the use of Hazard Communication (HazCom) Standard pictograms, older chemicals or safety information may still include these labels. You have probably seen the fire diamond before, possibly even while driving! It is often present on the side of trucks that are used to ship chemicals. A summary of this system is available at: https://www.acs.org/content/acs/en/chemical-safety/basics/nfpa-hazard-identification.html.

HazCom pictograms were implemented in 2015. SDSs now include these pictograms. There are nine pictograms designed to communicate specific chemical safety concerns. Any new chemical purchased is labeled with all pictograms relevant to its safety information. The pictograms may be viewed at: https://www.osha.gov/Publications/HazComm_QuickCard_Pictogram.html.

For this week's laboratory, you will work with your team to access the SDS, fire diamond, and HazCom symbols for your assigned chemicals. Your team must fill out the attached sheets for each chemical and the final summary sheet.

Safety data sheet information

Name of chemical:
Source for SDS (provide full URL):
Other names by which the chemical is known:
Appearance of chemical:
Physical state (solid, liquid, or gas):
Taste/Odor:
Boiling point:
Melting point:
Incompatible materials:
Health hazards (include symptoms, target organs):
What should you do if you spill the chemical on your skin?
What should you do if you inhale the chemical?
What should you do if you ingest the chemical?
Is the chemical a fire hazard?
What types of safety equipment should be used when handling the chemical?

NFPA (fire diamond) information

Source (provide full URL):
Health Hazard Score: What does this score mean?
Flammability Score: What does this score mean?

Instability Score: What does this score mean?
Specific Hazards (include symbol and meaning):

Hazard communication standard information
Source (provide full URL):

Which of the nine pictograms (Health Hazard, Flame, Exclamation Mark, Gas Cylinder, Corrosion, Exploding Bomb, Flame Over Circle, Environment, and Skull and Crossbones) are assigned to this chemical?
Use the Pictogram Quick Card provided at the OSHA website https://www.osha.gov/Publications/HazComm_QuickCard_Pictogram.html to explain what each of the pictograms for your chemical tell you.

Summary Sheet: Complete *after* your group has finished the individual worksheets for all three of your assigned chemicals.

1. List the names of the chemicals that your group investigated.
2. Why is it important to know what materials are incompatible with a chemical?
3. Which of the assigned chemicals is the most hazardous if you spill it on your skin? Why?
4. Which of the assigned chemicals is the most hazardous through inhalation? How do you know?
5. Which of the assigned chemicals is the most hazardous if ingested? How do you know?
6. When working with chemicals, it is important to know both the hazards of a chemical and what routes of exposure are likely to lead to the hazards. Which two of the three routes of exposure in questions 3-5 is it absolutely critical that you know for every chemical you handle in the laboratory? Why is the third route of exposure less important for the organic chemistry laboratory?

3.8.2 Assignment sheet for fall end-of-semester project
Project: designing a chemical reaction
You will be working individually on this project. Your task is to design and carry out a chemical reaction. You will choose a starting material from the list at the end of this handout and will find a procedure for a reaction that you can do using this starting material. The procedure can come from an organic chemistry laboratory manual (your instructor has several available outside her office) or from a published journal article. You should not search for a laboratory that uses the specific chemical you have chosen. Instead, you should search for procedures that use chemicals in the same family as your starting material, so that it is reasonable to expect that it will work. You will then adjust the procedure for your starting material. You may not use a procedure that has already been used in the laboratory this semester. You will have up to 5 mL of starting material to work with. You are not required to use the full amount.

You will have two weeks to prepare your plan. Your plan must include the elements outlined below. Laboratory work may be done during the last two weeks of class.

Components of plan

- Choice of starting material to be used
- Sketch of the chemical reaction
- A complete list of all chemicals (including amounts) and glassware needed for your reaction.
- A complete procedure for carrying out the reaction, including purification and analysis. Do not use a method of analysis that we have not yet learned in the laboratory. Be aware that you will need to alter amounts of chemicals in procedures and will need to look up literature values for your expected products.
- References for any book or journal article that you used in designing your procedure, including literature values for products. If you used a laboratory manual that is not part of Dr. Workman's collection, also submit a photocopy of the laboratory procedure.

Starting material options

1-octanol
cyclohexyl bromide
1-octene
cyclohexene

3.8.3 Chemical literature assignment: Navigating the ACS publications page

Chemical literature assignment

Important! You will need to complete this section using a campus computer or log in through the library's proxy page.

Go to the website http://www.pubs.acs.org. This is the ACS's publication page. Click on the ACS Journals option in the menu and choose the Journal of Organic Chemistry (JOC).

1. On the JOC site, click on the tab that says "Current Issue." Choose an article from the current issue. Provide a full, correct bibliographic citation for the article. Refer to the ACS style manual available on D2L.
2. Click on the Articles ASAP (As Soon As Publishable) tab. Articles in this section are ready for readers but have not yet been published in the print version of the journal. Choose an article from the ASAP articles. Provide a full, correct bibliographic citation for the article.

Return to the main ACS publication page. For this part of the assignment, we want to find a procedure to oxidize an alcohol. Use the search box in the upper right of the screen. Input the search terms *alcohol oxidation*.

3. How many results are obtained?

4. Scan the results on the first page. From the information provided, how many of these results are likely to contain information that is relevant to the reaction you want to perform?

This search produced a very large number of results. It is possible that reading some of these articles will lead you to a good procedure, but with this many results it is probably a good idea to try to narrow the field by continuing to refine your search terms.

To refine your search terms, you should think carefully about what you want to do. If, for example, the alcohol that you want to oxidize is a *primary* alcohol, adding this modifier to the search could help pare down the results. Modify your search to look specifically for *primary alcohol oxidation*.

5. How many results are obtained?

While this is still a large number of results, it is significantly reduced from the previous search. Scanning the first page, you can see several articles in which a primary alcohol has been oxidized. These are more likely to be useful for us than the general results we obtained in the previous search. If we did not care what type of reagent we used, we could easily find a procedure that we could try. Modify your search terms one more time, adding the reagent PCC (pyridinium chlorochromate) to the terms used in the last search.

6. How many results are obtained?

Scan through the search results. You will notice that the first several articles are about oxidizing primary alcohols using reagents that are *not* PCC. Why did this happen? PCC is an older reagent. Many of the articles you see feature newer reagents which the articles compare to PCC. These articles are returned in this search because there is a good possibility that somebody interested in PCC oxidation of primary alcohols would also be interested in the articles about the new reagents.

Click on the *Advanced Search* button in the upper right hand corner of the search box. Type the search term *alcohol oxidation* into the first text box. Click on the Modify Selection button under the heading Content Type. Select the Journal of Organic Chemistry and click on the Update button on the top left. Hit the search button to repeat your search looking only in JOC.

7. How many results are obtained?

Click on the *Citation* tab in the search box on the upper right of the page. Use the citation search function to find the article with the citation below:

Liu, R.; Liang, X.; Dong, C.; Hu, X. *J. Am. Chem. Soc.* **2004**, *126*, 4112.

On the right-hand side of the page, you will see a box titled *Article Options*. The three options on the left of the box bring you to the article. The options on the right-hand side offer links to specific portions of the article, figures, and additional information. To complete the questions below, you will need the full article and the information available by following the *Supporting Info* link. When you click on this link, it will take you to a new page. You must click the link for the .pdf to see the supporting information. Among other things, the supporting information for this article includes the actual procedure the researchers used. You will need it to answer question 9.

8. What reaction was performed? (explain in general terms, you do not need to talk about each reaction individually).
9. What was the percent yield of the reaction of cyclohexanol to produce cyclohexanone?
10. How much of each of the following reagents were used for the reaction in question 9?
    a. cyclohexanol
    b. TEMPO
    c. Br_2
    d. $NaNO_2$
11. Assume you want to perform this reaction, but you want to start with 1.5 g of cyclohexanol. Write out the procedure you would need to follow. You should write the procedure as you would in your laboratory notebook.

3.8.4 Assignment sheet for Project 1: Greening a published experiment, honors section

Honors organic chemistry laboratory project

The laboratory project for Honors Organic Chemistry is an individual exploration in green chemistry. You will attempt to improve upon a published organic chemistry laboratory by applying principles of green chemistry to alter the procedure to make the laboratory greener. You will choose an experiment from the CHE 361 laboratory textbook, Modern Projects and Experiments in Organic Chemistry: Miniscale and Standard Taper Microscale. You will carry out the reaction once as written and then twice more using your proposed modifications. After collecting all data, you will apply green chemistry metrics to analyze each reaction. You will report the results of your investigation by giving a 10–15 min PowerPoint presentation.

Assignment

Choose an experiment from the CHE 361 laboratory textbook, Modern Projects and Experiments in Organic Chemistry: Miniscale and Standard Taper Microscale 2nd Edition by Mohrig, Hammond, Schatz, and Morrill, and decide which of the 12 Principles of Green Chemistry you would like to try to improve upon. You will be designing two modified reactions and must use two different types of changes, each focusing on a different green chemistry principle. For example, experimenting with a less hazardous reagent in one reaction and a greener solvent in the other is considered two different types of changes. Changing the solvent twice is not. You must use published journal articles as sources to modify your procedure. Once you have identified the changes you wish to make, you must submit your proposal to your instructor for approval. The required contents of your proposal are detailed below.

After your project has been approved, you will schedule laboratory time to complete your project. The outcomes of your project, including analytical data and green metric analysis, will be reported in your final presentation.

Timeline

The timeline for completion of the steps in this project is outlined below. Steps may be completed early.

> March 2: Choose an experiment from the textbook and submit for instructor approval
> March 19: Submit your project proposal for instructor approval and feedback
> April 1: Schedule laboratory time
> May 1: Laboratory work complete
> May 4: Submit a copy of your PowerPoint presentation to your instructor via email
> May 7: Presentations

Proposal contents

Your proposal should contain the name and page number of the original reaction. For each of your modified reactions, provide the following information:

- A written description of the intended change and how, if successful, it will improve on the greenness of the reaction.
- A sketch of the reaction, including the change.
- A complete written procedure for the reaction.
- A complete list of all chemicals, including amounts, needed for the reaction and any special glassware or equipment. This is important in determining whether we have or can acquire the needed materials. If a piece of glassware is something that you have already used in CHE 351 or CHE 361, it does not need to be included. If you have not used it, be sure to include it on the list. If you have no idea what it is, it would be wise to speak with your instructor before submitting your proposal.
- A complete list of references used to design your procedures. Journal articles should be referenced using the citation style below:
  Author Lastname, Author Initials; Second Author Last name, Second Author Initials. *Journal.* **Year**, *Volume,* first page number.

Please note and make use of appropriate italicizing and bolding as shown above.

Presentation contents

Your presentation should contain the following elements:

- Project title
- Description of the original reaction, including a drawing of the reaction
- Results of the original reaction (yield, purity, analytical data)
- Description of each modified reaction, including a drawing of the reaction
- Explanation of the goal of each modification
- A list of which of the 12 Principles you intended to improve upon
- Results of the modified reaction (yield, purity, analytical data)
- Green metrics calculations (atom economy, experimental atom economy, E factor, and effective mass yield) for each of the three reactions
- Conclusions
- References (use format above)

3.8.5 Assignment sheet for Project 2

Honors organic chemistry laboratory project

The laboratory project for Honors Organic Chemistry is an individual exploration in green chemistry. You will attempt to develop a laboratory experiment based on an alternative feedstock. You may choose from one of two options: use of a plant-based source (renewable feedstock) or use of a postconsumer product (recycled feedstock). Procedures for working with both types of materials are readily available. After determining which project option you would like to pursue, you will need to identify a specific feedstock to use for your project and find a procedure from a journal article isolating a specific molecule using that source. You will analyze your product to verify that you have obtained the correct molecule and demonstrate the molecule's availability to react by using it to carry out another chemical reaction.

Once you have identified the changes you wish to make, you must submit your proposal to your instructor for approval. The required contents of your proposal are detailed below. After your project has been approved, you will schedule laboratory time to complete your project. The outcomes of your project, including analytical data and green metric analysis will be reported in your final presentation.

Timeline

The timeline for completion of the steps in this project is outlined below. Steps may be completed early.

Due date for each activity

March 8: Choose a project and identify the feedstock you would like to focus on
March 22: Submit your project proposal for instructor approval and feedback
April 5: Laboratory work
May 3: Laboratory work complete

May 7: Submit a copy of your PowerPoint presentation to your instructor via email

May 10: Presentations

Proposal contents

Your proposal (due no later than March 29) must include the following:

- Identification of the feedstock you intend to use and the organic chemical you expect to isolate from the feedstock.
- A one-paragraph description of the procedure or reaction you plan to use to isolate an organic chemical from your chosen feedstock and of the reaction you plan to perform with your isolated chemical.
- A complete, typed procedure detailing the steps you will take to isolate your chemical.
- Plans to verify the identity of the chemical isolated (boiling point, IR, NMR, MS, etc.).
- A complete written procedure for the reaction you will perform with your chemical.
- Either a print copy of the sources you used for your procedures or complete citations including a full URL for an online procedure.
- A complete list of all chemicals needed, including amounts, and any special glassware or equipment. This is important in determining whether we have or can acquire the needed materials. If a piece of glassware is something that we have already used in CHE 351 or CHE 361, it does not need to be included. If we have not used it, be sure to include it on the list. If you have no idea what it is, it would be wise to speak with your instructor before submitting your proposal.
- A complete list of references used to design your procedures.

Presentation contents

Your presentation should be prepared in PowerPoint and contain the following elements:

- Project title
- Description of the feedstock used, how it is obtained, what chemical(s) can be isolated or directly produced from it.
- Results of the isolation procedure or reaction (yield, instrumental data, physical data)
- Description of the reaction you are performing with your isolated chemical, including a drawing of the reaction
- Results of the reaction (yield, purity, analytical data)
- Conclusions
- Suggestions for future work (what would you try next or do differently if you were to continue this project?)
- References

Acknowledgments

I would like to thank Scott Sargent, who served in both the role of laboratory manager and of Chemical Hygiene Officer at the University of Wisconsin–Stevens Point at Wausau. His assistance in reviewing proposals, obtaining materials, and helping to supervise the honors students' laboratory work have been invaluable to the success of these projects.

References

1. *ACS Guidelines and Evaluation Procedures for Bachelor's Degree Programs;* American Chemical Society: Washington, DC, 2015. https://www.acs.org/content/dam/acsorg/about/governance/committees/training/2015-acs-guidelines-for-bachelors-degree-programs.pdf.
2. Murray, D. H.; Obare, S.; Hageman, J. Early Research: A Strategy for Inclusion and Student Success. In *The Power and Promise of Early Research;* Murray, D. H., Obare, S. O., Hageman, J. H., Eds.; American Chemical Society: Washington, DC, 2016; pp 1–32.
3. Osborn, J. M.; Karukstis, K. K. The Benefits of Undergraduate Research, Scholarship, and Creative Activity. In *Broadening Participation in Undergraduate Research: Fostering Excellence and Enhancing the Impact;* Boyd, M., Wesemann, J., Eds.; Council on Undergraduate Research: Washington, DC, 2009; pp 41–53.
4. Weaver, G. C.; Wink, D. J.; Varma-Nelson, P.; Lytle, F. E. Integrating Research Into the Curriculum: Using Laboratory Modules to Engage Faculty and Students. In *Broadening Participation in Undergraduate Research: Fostering Excellence and Enhancing the Impact;* Boyd, M., Wesemann, J., Eds.; Council on Undergraduate Research: Washington, DC, 2009; pp 109–120.
5. Silverberg, L. J.; Tierney, J.; Cannon, K. C. Research at Predominantly Two-year Campuses of Penn State. In *The Power and Promise of Early Research;* Murray, D. H., Obare, S. O., Hageman, J. H., Eds.; American Chemical Society: Washington, DC, 2016; pp 83–118.
6. National Academies of Sciences, Engineering, and Medicine. *Integrating Discovery-Based Research into the Undergraduate Curriculum: Report of a Convocation;* National Academies Press: Washington, DC, 2015. http://www.nap.edu/catalog/21851/integrating-discovery-based-research-into-the-undergraduate-curriculum-report-of.
7. Hewlett, J. A. Undergraduate Research at the Community College: Barriers and Opportunities. In *The Power and Promise of Early Research;* Murray, D. H., Obare, S. O., Hageman, J. H., Eds.; American Chemical Society: Washington, DC, 2016; pp 137–151.
8. Fakayode, S. O.; Yakubu, M.; Adeyeye, O. M.; Pollard, D. A.; Mohammed, A. K. Promoting Undergraduate STEM Education at a Historically Black College and University Through Research Experience. *J. Chem. Educ.* **2014,** *91,* 662–665.
9. President's Council of Advisors on Science and Technology. Engage to Excel: Producing One Million Additional College Graduates With Degrees in Science, Technology, Engineering, and Mathematics, https://obamawhitehouse.archives.gov/sites/default/files/microsites/ostp/pcast-engage-to-excel-final_2-25-12.pdf.

10. Laredo, T. Changing the First-year Chemistry Laboratory Manual to Implement a Problem-based Approach That Improves Student Engagement. *J. Chem. Educ.* **2013,** *90,* 1151−1154.

11. Mataka, L. M.; Kowalske, M. G. The Influence of PBL on Students' Self-efficacy Beliefs in Chemistry. *Chem. Educ. Res. Pract.* **2015,** *16,* 929−938.

12. Schoffstall, A. M.; Gaddis, B. A. Incorporating Guided-Inquiry Learning Into the Organic Chemistry Laboratory. *J. Chem. Educ.* **2007,** *84,* 848−851.

13. Hammond, C. N.; Mohrig, J. R.; Colby, D. A. On the Successful Use of Inquiry-driven Experiments in the Organic Chemistry Laboratory. *J. Chem. Educ.* **2007,** *84,* 992−998.

14. Horowitz, G. The State of Organic Teaching Laboratories. *J. Chem. Educ.* **2007,** *84,* 346−353.

15. Domin, D. S. A Review of Laboratory Instruction Styles. *J. Chem. Educ.* **1999,** *76,* 543−547.

16. Powell, N. L.; Harmon, B. B. Course-Embedded Undergraduate Research Experiences: The Power of Strategic Course Design. In *The Power and Promise of Early Research;* Murray, D. H., Obare, S. O., Hageman, J. H., Eds.; American Chemical Society: Washington, DC, 2016; pp 119−136.

17. Tomasik, J. H.; Cottone, K. E.; Heethuis, M. T.; Mueller, A. Development and Preliminary Impacts of the Implementation of an Authentic Research-Based Experiment in General Chemistry. *J. Chem. Educ.* **2013,** *90,* 1155−1161.

18. Ghanem, E.; Long, S. R.; Rodenbusch, S. E.; Shear, R. I.; Beckham, J. T.; Procko, K.; DePue, L.; Stevenson, K. J.; Robertus, J. D.; Martin, S.; Holliday, B.; Jones, R. A.; Anslyn, E. V.; Simmons, S. L. Teaching Through Research: Alignment of Core Chemistry Competencies and Skills Within a Multidisciplinary Research Framework. *J. Chem. Educ.* **2018,** *95,* 248−258.

19. Sandi-Urena, S.; Cooper, M.; Stevens, R. Effect of Cooperative Problem-Based Lab Instruction on Metacognition and Problem-Solving Skills. *J. Chem. Educ.* **2012,** *89,* 700−706.

20. Kerr, M. A.; Yan, F. Incorporating Course-Based Undergraduate Research Experiences Into Analytical Chemistry Laboratory Curricula. *J. Chem. Educ.* **2016,** *93,* 658−662.

21. Winkelman, K.; Baloga, M.; Marcinkowski, T.; Giannoulis, C.; Anquandah, G.; Cohen, P. Improving Students' Inquiry Skills and Self-efficacy Through Research-Inspired Modules in the General Chemistry Laboratory. *J. Chem. Educ.* **2015,** *92,* 247−255.

22. Kirchhoff, M.; Ryan, M. A. *Greener Approaches to Undergraduate Chemistry Experiments;* American Chemical Society: Washington, DC, 2002.

23. Doxsee, K. M.; Hutchison, J. E. *Green Organic Chemistry: Strategies, Tools, and Laboratory Experiments;* Brooks/Cole: Pacific Grove, CA, 2004.

24. Roesky, H. W., Kennepohl, D. K., Eds. *Experiments in Green and Sustainable Chemistry;* Wiley-VCH: Weinheim, Germany, 2009.

25. Bastin, L. D.; Gerhart, K. A Greener Organic Chemistry Course Involving Student Input and Design. In *Green Chemistry Experiments in Undergraduate Laboratories;* Fahey, J. T., Maelia, L. E., Eds.; American Chemical Society: Washington, DC, 2016; pp 55−69.

26. Anastas, P. T.; Warner, J. C. *Green Chemistry: Theory and Practice;* Oxford University Press: New York, NY, 1998.

27. Duarte, R. C. C.; Ribeiro, M. G. T. C.; Machado, A. A. S. C. Using Green Star Metrics to Optimize the Greenness of Literature Protocols for Syntheses. *J. Chem. Educ.* **2015,** *92,* 1024−1034.
28. Mercer, S. M.; Andraos, J.; Jessop, P. G. Choosing the Greenest Synthesis: A Multivariate Metric Green Chemistry Exercise. *J. Chem. Educ.* **2012,** *89,* 215−220.
29. Haley, R. A.; Ringo, J. M.; Hopgood, H.; Denlinger, K. L.; Das, A.; Waddell, D. C. Graduate Student Designed and Delivered: An Upper-Level Online Course for Undergraduates in Green Chemistry and Sustainability. *J. Chem. Educ.* **2018,** *95,* 560−569.
30. Bennet, G. D. Green Chemistry in Undergraduate Research. *CUR Quarterly* **2008,** *28,* 40−42.
31. Purcell, S. C.; Pande, P.; Lin, Y.; Rivera, E. J.; Paw U, L.; Smallwood, L. M.; Kerstiens, G. A.; Armstrong, L. B.; Robak, M. T.; Baranger, A. M.; Douskey, M. C. Extraction and Antibacterial Properties of Thyme Leaf Extracts: Authentic Practice of Green Chemistry. *J. Chem. Educ.* **2016,** *93,* 1422−1427.
32. Dintzner, M. R.; Maresh, J. J.; Kinzie, C. R.; Arena, A. F.; Speltz, T. A Research-Based Undergraduate Organic Laboratory Project: Investigation of a One-Pot, Multicomponent, Environmentally Friendly Prins-Friedel-Crafts-Type Reaction. *J. Chem. Educ.* **2012,** *89,* 265−267.
33. Mak, K. K. W.; Siu, J.; Lai, Y. M.; Chan, P. Mannich Reactions in Room Temperature Ionic Liquids (RTILs): An Advanced Undergraduate Project of Green Chemistry and Structural Elucidation. *J. Chem. Educ.* **2006,** *83,* 943−946.
34. Slade, M. C.; Raker, J. R.; Kobilka, B.; Pohl, N. L. B. A Research Module for the Organic Chemistry Laboratory: Multistep Synthesis of a Fluorous dye Molecule. *J. Chem. Educ.* **2014,** *91,* 126−130.
35. Gross, E. M.; Williams, S. H.; Williams, E.; Dobberpuhl, D. A.; Fujita, J. Synthesis and Characterization of Biodiesel From Used Cooking Oil: A Problem-Based Green Chemistry Laboratory Experiment. In *Green Chemistry Experiments in Undergraduate Laboratories;* Fahey, J. T., Maelia, L. E., Eds.; American Chemical Society: Washington, DC, 2016; pp 71−92.
36. Lee, N. E.; Gurney, R.; Soltzberg, L. Using Green Chemistry Principles as a Framework to Incorporate Research Into the Organic Laboratory Curriculum. *J. Chem. Educ.* **2014,** *91,* 1001−1008.
37. Edgar, L. J. G.; Koroluk, K. J.; Golmakani, M.; Dicks, A. P. Green Chemistry Decision-Making in an Upper-Level Undergraduate Organic Laboratory. *J. Chem. Educ.* **2014,** *91,* 1040−1043.
38. Graham, K. J.; Jones, T. N.; Schaller, C. P.; McIntee, E. J. Implementing a Student-Designed Green Chemistry Laboratory Project in Organic Chemistry. *J. Chem. Educ.* **2014,** *91,* 1895−1900.
39. *Creating Safety Cultures in Academic Institutions: A Report of the Safety Culture Task Force of the ACS Committee on Chemical Safety,* 1st ed.; American Chemical Society: Washington, DC, 2012.
40. Huston, E. M.; Milligan, J. A.; Powell, J. R.; Smith, A. M.; Neal, D.; Duval, K. M.; DiNardo, M. A.; Stoddard, C.; Bell, P. A.; Berning, A. W.; Wipf, P.; Bandik, G. C. Development of an Undergraduate Course in Chemical Laboratory Safety Through an Academic/Industrial Collaboration. *J. Chem. Educ.* **2018,** *95,* 577−583.
41. Kennedy, S.; Palmer, J. Teaching Safety: 1000 Students at a Time. *J. Chem. Health Saf.* **2011,** *18,* 26−31.

42. Alaimo, P. J.; Langenhan, J. M.; Tanner, M. J.; Ferrenberg, S. M. Safety Teams: An Approach to Engage Students in Laboratory Safety. *J. Chem. Educ.* **2010,** *87,* 856−861.

43. Wright, S. M. Introducing Safety Topics Using a Student-Centered Approach. *J. Chem. Educ.* **2005,** *82,* 1519−1520.

44. Miller, G. J.; Heideman, S. A.; Greenbowe, T. J. Introducing Proper Chemical Hygiene and Safety in the General Chemistry Curriculum. *J. Chem. Educ.* **2000,** *77,* 1185−1187.

45. Stuart, R. B.; McEwen, L. R. The Safety "Use Case": Co-developing Chemical Information Management and Laboratory Safety Skills. *J. Chem. Educ.* **2016,** *93,* 516−526.

46. Bradley, S. Integrating Safety Into the Undergraduate Chemistry Curriculum. *J. Chem. Health Saf.* **2011,** *18,* 4−10.

47. Sigmann, S. Incorporating the New American Chemical Society Safety Guidelines Into an Undergraduate Chemistry Program. *J. Chem. Health Saf.* **2011,** *18,* 11−15.

48. Crockett, J. M. Laboratory Safety for Undergraduates. *J. Chem. Health Saf.* **2011,** *18,* 16−25.

49. Mohrig, J. R.; Hammond, C. N.; Schatz, P. F.; Morrill, T. C. *Modern Projects and Experiments in Organic Chemistry: Miniscale and Standard Taper Microscale,* 2nd ed.; W.H. Freeman: New York, NY, 2003.

50. McKenzie, L. C.; Thompson, J. E.; Sullivan, R.; Hutchison, J. E. Green Chemical Processing in the Teaching Laboratory: A Convenient Liquid CO_2 Extraction of Natural Products. *Green Chem.* **2004,** *6,* 355−358.

51. Esteb, J. J.; Magers, J. R.; McNulty, L.; Morgan, P.; Wilson, A. M. A Simple S_N2 Reaction for the Undergraduate Laboratory. *J. Chem. Educ.* **2009,** *86,* 850−852.

52. Taber, R. L.; Champion, W. C. Dehydration of 2-methylcyclohexanol. *J. Chem. Educ.* **1967,** *44,* 620.

53. Gawalt, E. S.; Adams, B. A Chemical Information Literacy Program for First-Year Students. *J. Chem. Educ.* **2011,** *88,* 402−407.

54. Bruehl, M.; Pan, D.; Ferrer-Vinent, I. J. Demystifying the Chemistry Literature: Building Information Literacy in First-Year Chemistry Students Through Student-Centered Learning and Experiment Design. *J. Chem. Educ.* **2015,** *92,* 52−57.

55. Forest, K.; Rayne, S. Incorporating Primary Literature Summary Projects Into a First-Year Chemistry Curriculum. *J. Chem. Educ.* **2009,** *86,* 592−594.

56. Locknar, A.; Mitchell, R.; Rankin, J.; Sadoway, D. R. Integration of Information Literacy Components Into a Large First-Year Lecture-Based Chemistry Course. *J. Chem. Educ.* **2012,** *89,* 487−491.

57. Jensen, D., Jr.; Narske, R. Beyond Chemical Literature: Developing Skills for Chemical Research Literacy. *J. Chem. Educ.* **2010,** *87,* 700−702.

58. Yeagley, A. A.; Porter, S. E. G.; Rhoten, M. C.; Topham, B. J. The Stepping Stone Approach to Teaching Chemical Information Skills. *J. Chem. Educ.* **2016,** *93,* 423−428.

59. Graham, K. J.; Schaller, C. P.; Jones, T. N. An Exercise to Coach Students on Literature Searching. *J. Chem. Educ.* **2015,** *92,* 124−126.

60. Swoger, B. J. M.; Helms, E. An Organic Exercise in Information Literacy Using SciFinder. *J. Chem. Educ.* **2015,** *92,* 668−671.

61. Baykoucheva, S.; Houck, J. D.; White, N. Integration of EndNote Online in Information Literacy Instruction Designed for Small and Large Chemistry Courses. *J. Chem. Educ.* **2016,** *93,* 470−476.

62. Russel, G. A.; Desmond, K. M. Directive Effects in Aliphatic Substitutions. XIX. Photobromination with N-Bromosuccinimide. *J. Am. Chem. Soc.* **1963,** *85,* 3139−3141.

63. Nelson, W. M. Choosing Solvents That Promote Green Chemistry. In *Green Chemical Syntheses and Processes;* Anastas, P. T., Heine, L. G., Williamson, T. C., Eds.; American Chemical Society: Washington, DC, 2000; pp 313−328.

64. Pavlis, R. R. An Easily Conducted Free Radical Substitution for Organic Chemistry Courses. *J. Chem. Educ.* **1982,** *59,* 658.

65. Henderson, R. K.; Jiménez-González, C.; Constable, D. J. C.; Alston, S. R.; Inglis, G. G. A.; Fisher, G.; Sherwood, J.; Binks, S. P.; Curzons, A. D. Expanding GSK's Solvent Selection Guide − Embedding Sustainability into Solvent Selection Starting at Medicinal Chemistry. *Green Chem.* **2011,** *13,* 854−862.

66. McKenzie, L. C.; Huffman, L. M.; Hutchison, J. E. The Evolution of a Green Chemistry Laboratory Experiment: Greener Brominations of Stilbene. *J. Chem. Educ.* **2005,** *82,* 306−310.

67. Mahmoud, M. E.; Abdou, A. E. H.; Ahmed, S. B. Conversion of Waste Styrofoam into Engineered Adsorbents for Efficient Removal of Cadmium, Lead and Mercury From Water. *ACS Sustain. Chem. Eng.* **2016,** *4,* 819−827.

68. Padmapriya, A. A.; Just, G.; Lewis, N. G. A New Method for the Esterification of Sulphonic Acids. *Synth. Commun.* **1985,** *15,* 1057−1062.

69. Alaimo, P. J.; Langenhan, J. M.; Suydam, I. T. Aligning the Undergraduate Organic Laboratory Experience With Professional Work: The Centrality of Reliable and Meaningful Data. *J. Chem. Educ.* **2014,** *91,* 2093−2098.

Microwave reactivity and energy efficiency in the undergraduate organic laboratory

Andrew P. Dicks, PhD

*Professor, Teaching Stream, Department of Chemistry, University of Toronto,
Toronto, Ontario, Canada*

4.1 Introduction

As a familiar appliance to many of us in domestic kitchens, microwave ovens have featured in chemistry laboratories for several decades. The more traditional ovens have morphed into dedicated reactors used for a myriad of purposes within different chemical fields. Microwave chemistry has been described as "hot, fast, and a tad mystical" *(1)* and is currently of interest to many academics in terms of research and teaching, as well as the pharmaceutical and chemical industries. This is largely due to a significant reduction in reaction time that is possible with microwave irradiation, compared with more classical heating methods using oil or water baths. Microwave energy is converted into heat when polarized molecules interact with one another in an attempt to align with the applied electromagnetic field. This means a reaction vessel does not have to be directly heated in a way that is required by conventional approaches, where heat is transferred by convection. An excellent, recent overview of general concepts relating to microwave-assisted synthesis has been written by Leadbeater *(2)*. In a microwave reactor, high temperatures are reached much more quickly than by utilizing a liquid heating bath, and (if using sealed vessels) it becomes feasible to attain conditions where the boiling point of a reaction solvent is exceeded ("superheating"). This is often described as a "pressure cooker effect." These factors logically extend to faster transformations often being observed using microwave heating. Implicit here is that a faster reaction *should* be a less energy-demanding one; therefore microwave chemistry often has the tag of being a "greener technology."

4.1.1 Microwave reactivity: A summary of advances in the research laboratory

A number of articles are highlighted in this section that will be of interest to educators wishing to learn about recent developments in microwave research.

Integrating Green and Sustainable Chemistry Principles into Education. https://doi.org/10.1016/B978-0-12-817418-0.00004-8
Copyright © 2019 Elsevier Inc. All rights reserved.

A long-standing debate in the area of microwave reactivity is the existence (or otherwise) of specific or "nonthermal" effects that contribute to substantial reaction rate enhancements. In 2013, after a decade of research in the area, Kappe et al. concluded that nonthermal microwave effects do not exist and that "microwave chemistry is not 'voodoo science'" *(3)*. In contrast, 2 years later, Dudley et al. reported in a *Chemical Science* perspective that the unique heating properties of microwaves create solution dynamics that cannot be rationalized in terms of current theory *(4)*. One explanation is that rapid heating from microwave irradiation could impact the frequency and type of molecular collisions, thereby enhancing the rate of reaction without influencing the temperature. More recently, Prieto and coworkers have illustrated how computational calculations can assist the justification of observed product selectivity in microwave reactions *(5)*, so that prediction of improvements becomes possible. Three separate textbook chapters discussing general applications of microwave-assisted organic synthesis have been written *(6—8)*, with more specific reviews written in the areas of heterogeneous catalysis *(9)*, nanomaterials *(10)*, and carbon—carbon bond forming cross-couplings *(11)*. The opportunity to perform microwave chemistry in benign reaction media (including water) has also been highlighted *(12,13)*.

4.1.2 An overview of microwave technology in teaching laboratories

It is not a purpose of this chapter to present the theory behind microwave chemistry as it relates to classroom and laboratory instruction, as this has been previously covered in the pedagogical literature. The concept of microwave technology in undergraduate laboratories was highlighted by Cresswell and Haswell in 2001 *(14)*, and by Zovinka and Stock in 2010 *(15)*. The latter article, entitled "Microwave Instruments: Green Machines for Green Chemistry?" makes a direct connection between microwave heating and "…the introduction of the ethics and the philosophy inherent in green chemistry." One of the 12 Principles of Green Chemistry is, of course, "design for energy efficiency" *(16)*, with an implication that the reduced heating times obtainable in a microwave oven lead to reactions that save energy. Microwave chemistry also lends itself to undertaking transformations in environmentally friendly solvents such as water, or sometimes in the absence of any solvent at all. Performing reactions under these conditions facilitates connections for students with other green chemistry principles, including those related to waste prevention, less hazardous synthesis, and benign solvents/auxiliaries *(16)*. A number of organic experiments that feature microwave heating and underscore these concepts have been reviewed *(17—19)*.

Since 2000, more than 50 laboratory activities have been published in the *Journal of Chemical Education* that incorporate microwave chemistry in one way or another. The impression may be that most of these have been devised for incorporation into organic curricula, but it is worth noting that important advances have been made in general/inorganic courses *(20—33)* and in analytical offerings *(34—36)*. An interesting development is the recent design of microwave experiments that showcase principles of nanochemistry *(37–42)*. From an organic perspective, the reader is directed to a didactic textbook chapter that summarizes microwave heating as a greener technology *(43)*. As well as reviewing some relevant examples from the teaching literature, this account gives some historical context and discusses microwave versus conventional heating methods. Most importantly, it provides a personal description of how the author developed various experiments herself, with advice for other instructors. As part of this section, some guidance is forthcoming regarding the purchase and operation of a laboratory-grade microwave reactor, including the major differences between monomode and multimode ovens. Monomode units have small cavities and feature sample irradiation at a single wavelength, whereas multimode ovens utilize multiple wavelengths of microwave energy within larger cavities (similar to domestic microwave ovens). The latter are the ones most often found in undergraduate laboratories due to their flexibility in the number of samples heated and the reaction scale. On the other hand, monomode ovens can promote particularly fast reactions so that the apparent disadvantage of running one sample at a time is negated (as introduced in Section 4.2.2).

4.2 Recent pedagogical experiments highlighting microwave-assisted organic chemistry

The student experiments highlighted in this section represent those published between 2011 and 2018, following the textbook chapter written by Baar *(43)*. They are largely organized according to the type of operative reaction mechanism as follows: (1) nucleophilic additions; (2) nucleophilic substitutions; (3) electrophilic substitutions; (4) eliminations; (5) cycloadditions; and (6) transition metal-catalyzed reactions. Two further categories included are multicomponent/multistep reactions and extractions. It is important to note that the vast majority of these experiments were designed to be performed in commercial microwave reactors, rather than domestic ovens. Indeed, the *Journal of Chemical Education* states the following policy in its 2018 author guidelines: "The *Journal* does not publish manuscripts that involve the use of domestic (i.e., kitchen) microwave ovens because such use is potentially hazardous and poses safety concerns" *(44)*. Although the actual reaction temperature is not reported in many of the experimental descriptions, an indication of the student product yield is included here for each reaction (where known), along with the microwave heating time. This generally refers to the so-called "fixed hold time" at a specific temperature and does not include minutes dedicated to either initial heating (the "ramp time") or oven cool-down.

4.2.1 Nucleophilic addition reactions

Russell and coworkers have described a laboratory curriculum designed to transition undergraduates from "cookbook chemistry" to several student-centered experiments framed around microwave heating *(45)*. One example is the reduction of acetophenone (Scheme 4.1) which proceeds in 30 s under microwave irradiation. As the same reaction takes place under reflux (10 min), students are divided into pairs and are able to compare different modes of heating and the merits of each technique. In comparison, reduction of nitrobenzene and four derivatives to the corresponding aniline products has been reported in a modified domestic microwave oven. The reductant used is NH_2NH_2 featuring Pd/C as a catalyst, with a typical reaction time of 2 min in ethanol ((46), Scheme 4.2).

The artificial sweetener known as dulcin ((4-ethoxyphenyl)urea) has been prepared by second semester organic undergraduates from the microwave-assisted reaction between p-phenetidine and urea (Scheme 4.3). The short reaction time (5 min at 125°C) compares favorably with that of at least 30 min under more conventional conditions. The key mechanistic step is the nucleophilic addition of p-phenetidine to isocyanic acid (H-N=C=O), which is generated in situ (47). In a similar spirit of requiring students to generate a "real-world relevant" compound, Anderluh designed a two-step synthesis of celecoxib (Celebrex, a nonsteroidal antiinflammatory drug) where the final reaction is undertaken under either conventional or microwave-assisted heating in a monomode reactor (Scheme 4.4). According to administered surveys, students found this experiment to be valuable in appreciating microwave chemistry as an important approach in organic synthesis (48). As previously highlighted (17), the undergraduate preparation of trans-4-methoxycinnamic acid via a microwave-heated Knoevenagel condensation has also been disclosed (49).

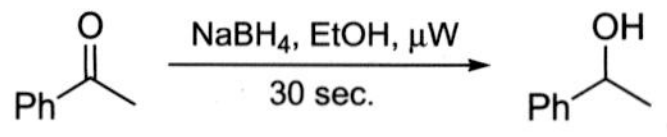

SCHEME 4.1

Reduction of acetophenone using sodium borohydride.

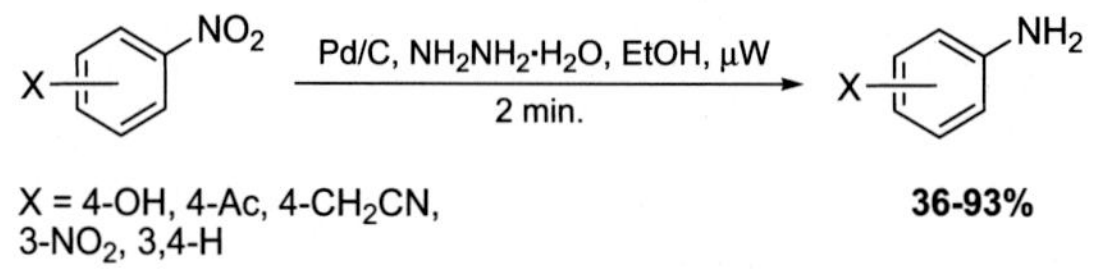

SCHEME 4.2

Aromatic nitro group reduction.

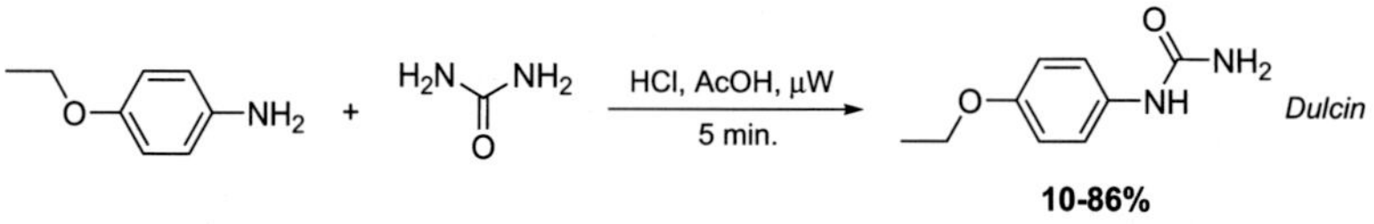

SCHEME 4.3

Preparation of an artificial sweetener.

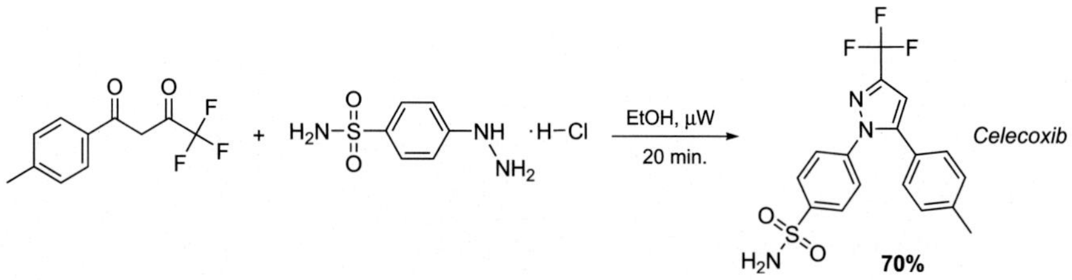

SCHEME 4.4

Synthesis of celecoxib, a nonsteroidal antiinflammatory drug.

4.2.2 Nucleophilic substitution reactions

4.2.2.1 S_N2 nucleophilic substitutions

Baar has implemented a research experience for the introductory organic laboratory where the class is charged with optimizing a Williamson ether synthesis (WES) of 2-fluorophenetole ((50), Scheme 4.5). This inquiry-based experiment focuses on varying particular parameters including monomode microwave heating temperature, time, and wattage. Students are also able to alter the reaction solvent (including its volume) and reagent equivalencies and are required to suggest improved conditions based on pooled class data. The same author previously outlined a more traditional microwave-assisted WES in the context of comparing monomode and multimode oven heating ((51), Scheme 4.6). A third microwave-heated WES was reported by Jensen et al. (52) where students react p-hydroxybenzoic acid with a primary alkyl bromide to generate one of the six different ethers that display liquid crystal phases (Scheme 4.7). Phase behavior is then characterized by melting point measurements, polarized optical microscopy, and differential scanning calorimetry.

Two further reactions proceeding via an S_N2 mechanism under microwave heating with nitrogen-based nucleophiles are worthy of attention. First, a high-yielding,

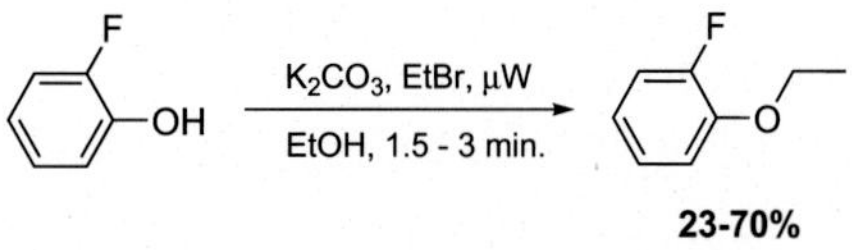

SCHEME 4.5

An inquiry-based Williamson ether synthesis.

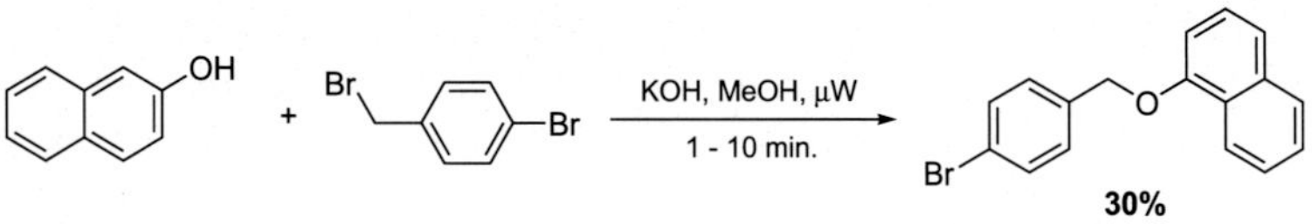

SCHEME 4.6

Williamson ether synthesis under monomode and multimode conditions.

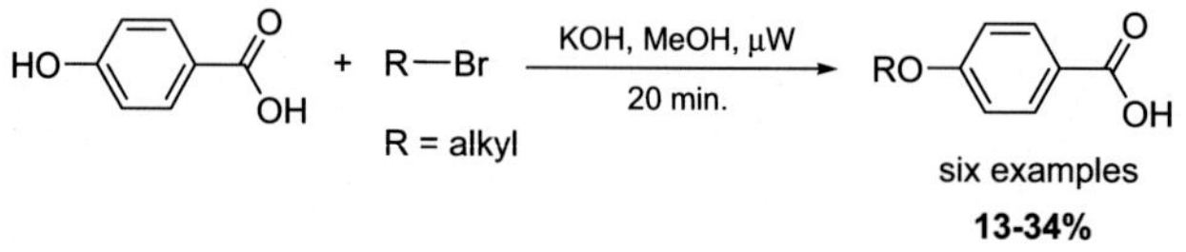

SCHEME 4.7

Preparation of liquid crystals from *p*-hydroxybenzoic acid.

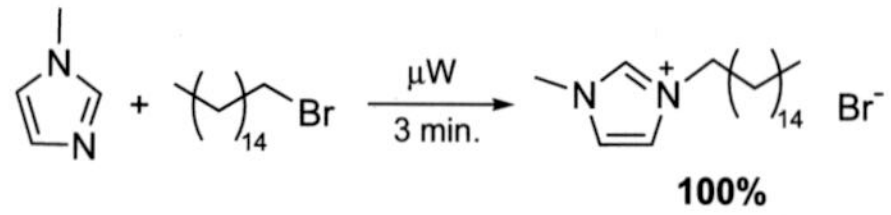

SCHEME 4.8

Solvent-free synthesis of an ionic liquid.

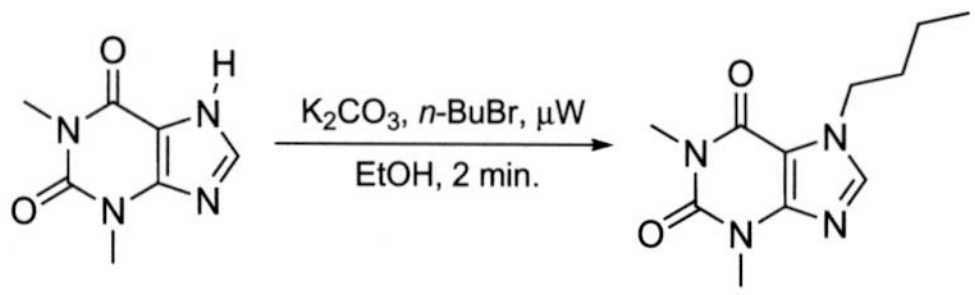

SCHEME 4.9

N-butylation of theophylline.

solvent-free ionic liquid synthesis has been presented by Hu and coworkers ((*53*), Scheme 4.8). The product is consequently used to manufacture an organic—inorganic mesoporous material in an experiment designed to introduce third-year undergraduates to research activities. In addition, as a second example from the curriculum of Russell et al. *(45)*, the *n*-butylation of theophylline presents an opportunity to adapt a literature procedure by changing the alkylating agent, base, and reaction scale (Scheme 4.9). Ultimately, small groups of students then attempt to optimize the *n*-butylation reaction by experimenting with different conditions. The microwave procedure provides a significant rate enhancement in this case compared to the corresponding room temperature alkylation.

4.2.2.2 Nucleophilic acyl substitutions

Fischer esterifications under microwave heating have been brought into the teaching laboratory by a number of groups. These procedures have proved popular with undergraduates due to the short reaction times and pleasant-smelling compounds that are generated. Lampkowski et al. have developed an esterification experiment using a CEM Discover microwave that utilizes acetic acid and four unknown alcohols to bring in a problem-solving aspect for students ((*54*), Scheme 4.10). In comparison, Baar used both a monomode oven (30 s heating) and a multimode oven (10 min heating) to establish a Fischer esterification protocol where students synthesize 11 esters varying in alkyl groups ((*51*), Scheme 4.11). A discovery-based microscale

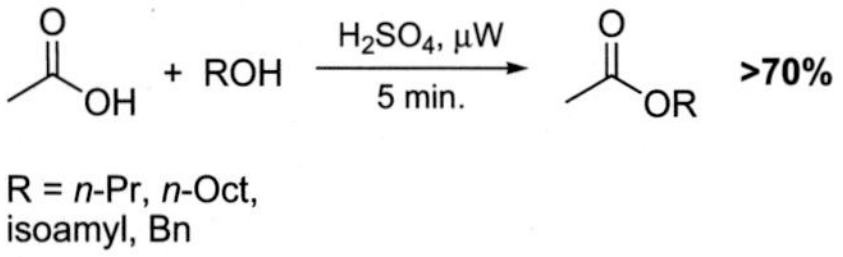

SCHEME 4.10

Fischer esterification of acetic acid to form four different esters.

Fischer esterification experiment that avoids an organic solvent and does not require product purification has been discussed previously *(17,55)*. In 2016, Fahey and co-workers published a microwave-assisted synthesis of aspirin (containing an ester functionality), following (1) extraction of methyl salicylate from over-the-counter pain creams; and (2) hydrolysis to salicylic acid *(56)*. The aspirin preparation step features use of naturally acidic fruit juices and soft drinks as catalysts rather than the more traditional sulfuric acid or phosphoric acid (Scheme 4.12).

Atom economical and solventless imide formation via microwave irradiation of aniline and succinic anhydride has been presented by Shell et al. *(57)*. The product (*N*-phenylsuccinimide) is formed in moderate yields on heating in a domestic microwave oven (Scheme 4.13) and is easily isolated and purified in a single organic laboratory period. Horta adopted a similar approach to design microwave-assisted Claisen and Dieckmann condensations for the undergraduate laboratory *((58)*,

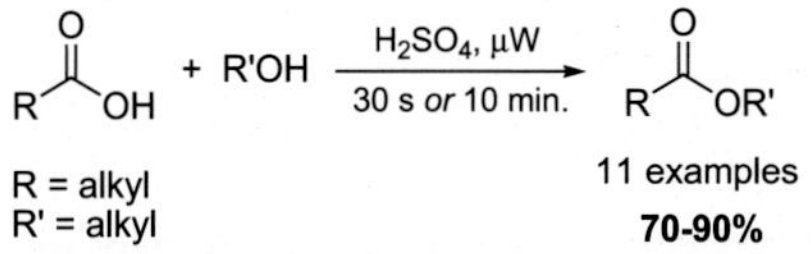

SCHEME 4.11

Ester formation under monomode and multimode conditions.

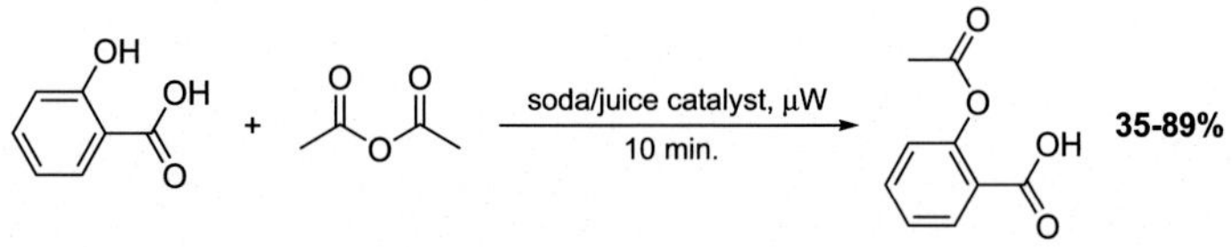

SCHEME 4.12

Preparation of aspirin from salicylic acid using naturally acidic consumer products.

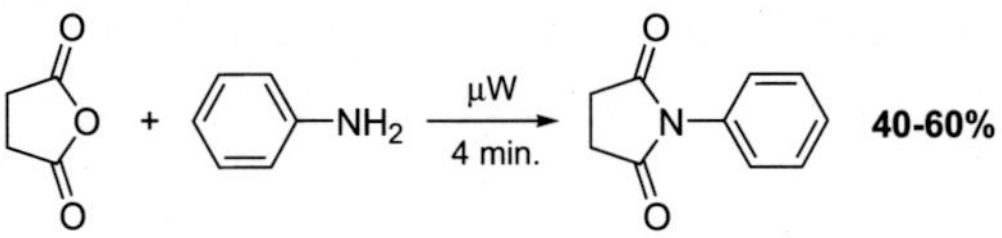

SCHEME 4.13

Solventless synthesis of *N*-phenylsuccinimide.

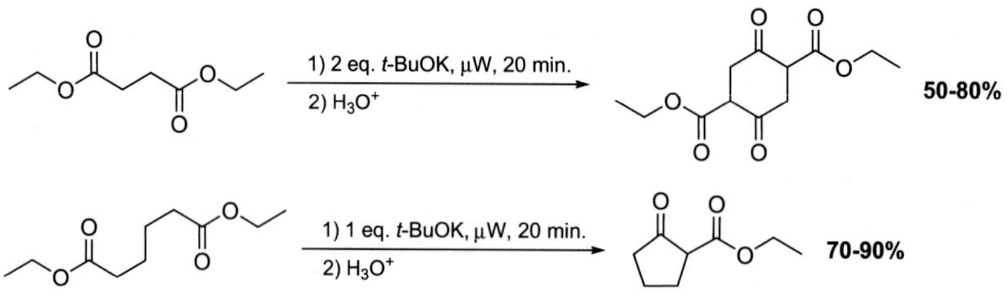

SCHEME 4.14

Claisen and Dieckmann condensation reactions.

Scheme 4.14). These two classical transformations are routinely discussed during the lecture component of introductory organic classes yet are rarely undertaken during laboratory time.

Basic hydrolysis of ester linkages via microwave heating has been employed to highlight aspects of green chemistry and sustainability. Polyethylene terephthalate (PET) waste from soft drink bottles can be depolymerized to generate terephthalic acid and ethylene glycol (commercial antifreeze) in the presence of a phase-transfer catalyst (PTC) such as benzalkonium chloride *(59)*. Students bring postconsumer PET bottles to the laboratory and grind them to a powder that is irradiated for 5 min in the presence of aqueous sodium hydroxide and a PTC (Scheme 4.15). This is in comparison with the conventional depolymerization route of heating with an oil bath for around 1 h. In a similar experiment, Pohl and coworkers have described the transesterification of triglycerides from plant oils to form biodiesel (in the presence of methanol or ethanol), or hydrolysis to form soap (in the presence of water) *((60)*, Scheme 4.16). Both of these reactions generate glycerol as a by-product. It is possible to use various oils (olive, soybean, and rapeseed) as the source of the triglycerides.

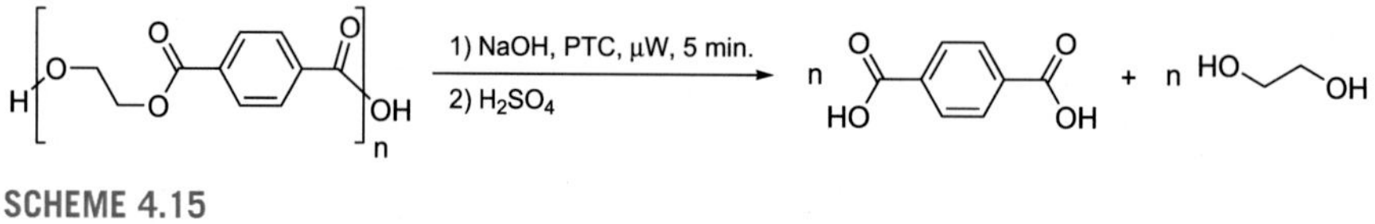

SCHEME 4.15

Depolymerization of polyethylene terephthalate under basic conditions.

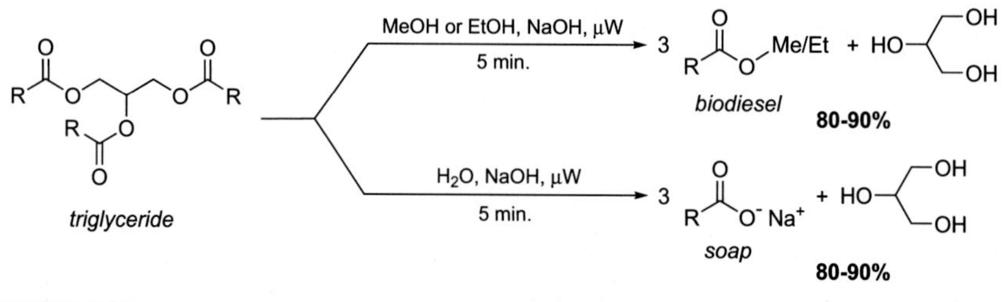

SCHEME 4.16

Transesterification and saponification of plant oil triglycerides.

4.2.2.3 A nucleophilic aromatic substitution

The microwave-irradiated substitution reactions of 1-bromo-2,4-dinitrobenzene with several nucleophiles have been reported ((*17,61*), Scheme 4.17). A comparison is made in terms of product yields between these processes and similar ones undertaken by conventional heating under reflux.

4.2.3 Electrophilic substitution reactions

Damkaci et al. have devised an inquiry-based, microwave-assisted Friedel–Crafts acylation of toluene using four different acid anhydrides that is suitable for incorporation into an introductory organic teaching laboratory ((*62*), Scheme 4.18). The anhydrides are provided to students as unknown reactants, and the product structures determined by various 1D NMR spectroscopic methods and IR spectroscopy. The aromatic ketone generated in each reaction is formed as a single regioisomer. A second electrophilic substitution microwave reaction (undergraduate nitration of phenol using $Cu(NO_3)_2$ in acetic acid to form a regioisomeric mixture (*63*)) has been reviewed previously (*17*).

4.2.4 Elimination reactions

An E1 elimination mechanism is often highlighted to students by performing the phosphoric acid-induced dehydration of cyclohexanol or a structurally related derivative. This reaction has been replaced by an operationally simple microwave reaction between α-tetralol and catalytic *p*-toluenesulfonic acid at Allegheny College ((*45*), Scheme 4.19). The presence of a chromophore in the substrate

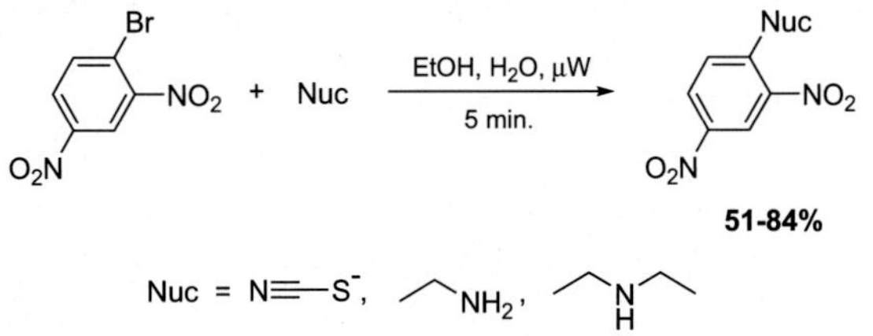

SCHEME 4.17

S_NAr reactions between three nucleophiles and 1-bromo-2,4-dinitrobenzene.

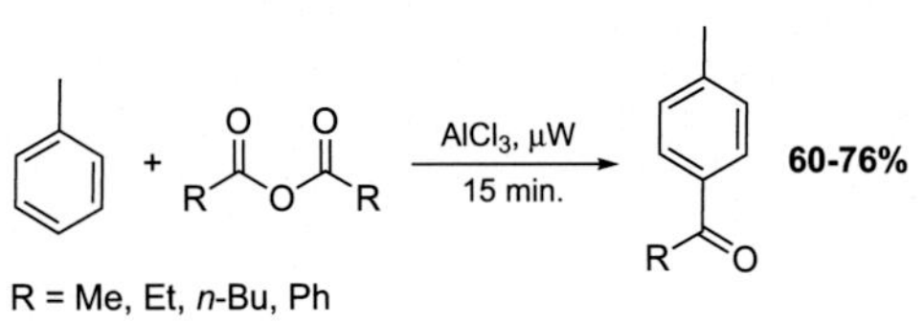

SCHEME 4.18

Friedel–Crafts acylation of toluene with acid anhydrides.

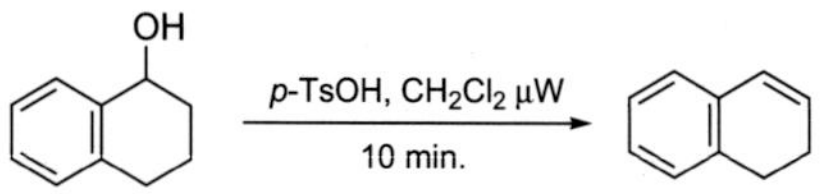

SCHEME 4.19

E1 dehydration of α-tetralol.

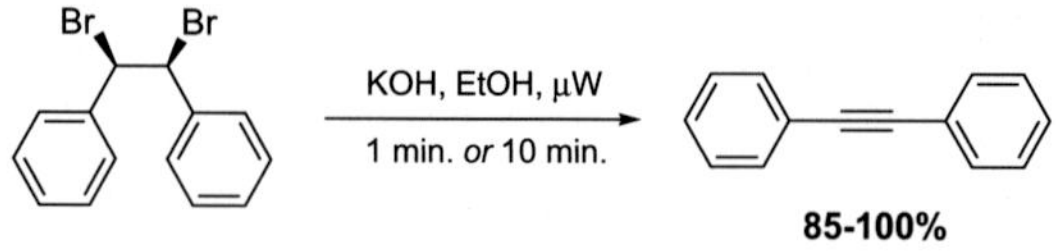

85–100%

SCHEME 4.20

E2 dehydrohalogenation of *meso*-1,2-dibromo-1,2-diphenylethane.

facilitates use of HPLC to follow reaction progress, with reagents being provided as solutions in dichloromethane. As an interesting contrast, the elimination of two molecules of HBr from *meso*-1,2-dibromo-1,2-diphenylethane to form diphenylacetylene is an exemplar of a microwave-assisted E2 mechanism under strongly basic conditions in either a monomode (1 min heating) or a multimode reactor (10 min heating) (*(51)*, Scheme 4.20).

4.2.5 Cycloaddition reactions

There are examples of Diels—Alder pericyclic reactions under microwave heating described in the recent pedagogical literature. One of these is the cycloaddition between 1,3-cyclohexadiene and *N*-phenylmaleimide to form the *endo* product isomer (*(51)*, Scheme 4.21). This reaction was found by students to proceed much faster in a monomode oven (30 s heating) compared with a multimode one (10 min heating), with identical yields being obtained. Saunders and coworkers outlined the undergraduate solvent-free synthesis of *cis*-norbornene-5,6-dicarboxylic anhydride as the *exo* isomer *(64)*. This can be achieved either directly from maleic anhydride and dicyclopentadiene or, as indicated in Scheme 4.22, from the corresponding *endo* isomer.

Very recently, some [3 + 2] cycloaddition reactions between a range of aryl nitriles and sodium azide in the presence of different catalysts (cerium(III) chloride,

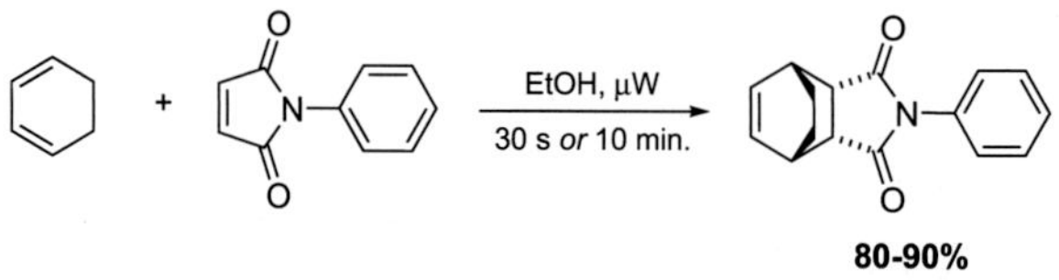

80–90%

SCHEME 4.21

Diels—Alder reaction between 1,3-cyclohexadiene and *N*-phenylmaleimide.

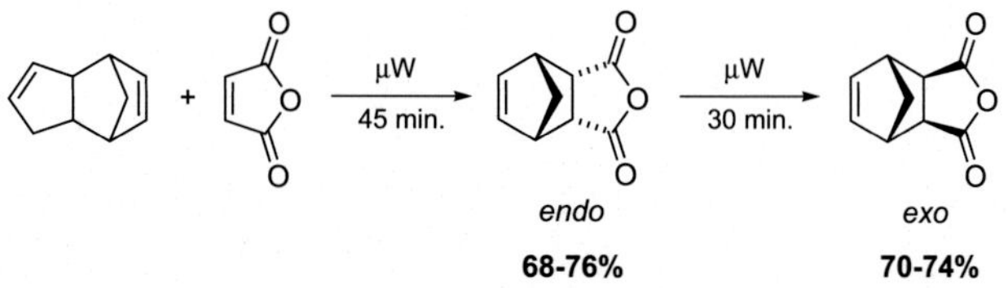

endo

68-76%

exo

70-74%

SCHEME 4.22

Preparation of *cis*-norbornene-*endo*-5,6-dicarboxylic anhydride.

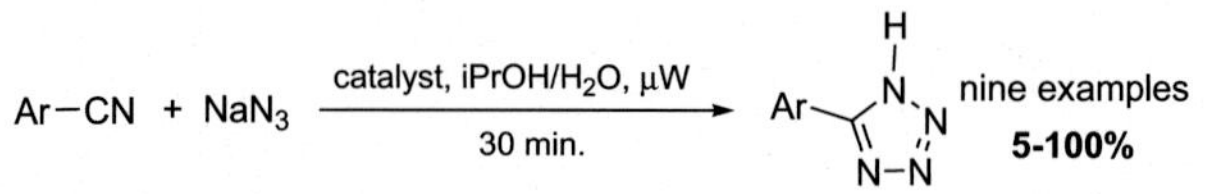

Ar—CN + NaN$_3$ $\xrightarrow[\text{30 min.}]{\text{catalyst, iPrOH/H}_2\text{O, μW}}$

nine examples

5-100%

SCHEME 4.23

[3 + 2] Cycloaddition reactions between aryl nitriles and sodium azide using different catalysts.

zinc(II) chloride or iron(III) chloride) have been disclosed *(65)*. These transformations occur via microwave heating for 30 min in a relatively green solvent combination (isopropanol/water, Scheme 4.23). Students are required to deduce the structure of their reaction product depending on their assigned aryl nitrile and collected spectroscopic data.

4.2.6 Transition metal-catalyzed reactions

Microwave-promoted palladium-catalyzed cross-couplings have been designed for the organic teaching laboratory in the form of Suzuki reactions. A three-week mini project published by Vargas et al. includes preparation of a biphenyl via coupling of phenylboronic acid and 4-iodoanisole with Pd/C as the catalyst (*(66)*, Scheme 4.24). This reaction is undertaken by students using an adapted domestic microwave oven and may also be incorporated into an undergraduate physical chemistry course. Soares and coworkers designed a similar experiment (Scheme 4.25) where water is used as the reaction solvent and different phenylboronic acid derivatives are utilized *(67)*. The phenylacetophenone products are important intermediates for synthesis of many biologically relevant compounds, including coumarins and flavonoids.

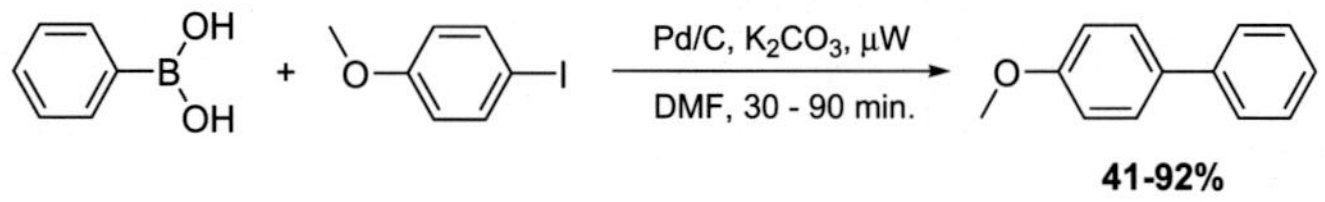

$\xrightarrow[\text{DMF, 30 - 90 min.}]{\text{Pd/C, K}_2\text{CO}_3\text{, μW}}$

41-92%

SCHEME 4.24

Suzuki coupling of phenylboronic acid and 4-iodoanisole.

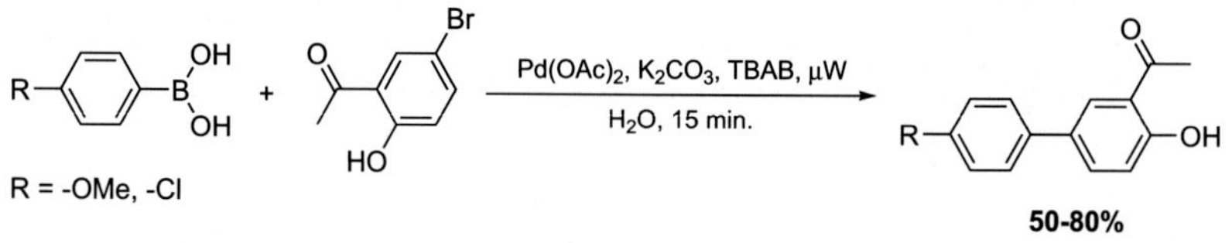

SCHEME 4.25

Suzuki coupling of phenylboronic acid derivatives and an aryl bromide.

4.2.7 Multicomponent and multistep reactions

The four-component Ugi reaction between an isocyanide, an amine, an aldehyde, and a carboxylic acid is an intrinsically green transformation as it exhibits exceptionally high atom economy. Ingold et al. have illustrated that such a reaction can be inserted into the organic curriculum and performed under "on-water" or solventless conditions *(68)*. It is possible for the bisamide product to be formed at room temperature, or in concert with ultrasound or microwave irradiation (Scheme 4.26). The authors additionally introduced the use of holistic metrics to allow students to consider the environmental impact of each set of reaction conditions and to assess which is/are greener *(69)*. Pohl et al. have focused on undergraduate-driven design of peptide mimetics where peptoid oligomers are formed in multistep processes *(70)*. Students are exposed to solid-phase synthesis in order to remove the requirement for intermediate isolations and purifications, and undertake microwave heating to impact reaction rates. The work of Slade et al. (multistep synthesis of a fluorous dye molecule, in which microwave chemistry is featured *(71)*) has previously been discussed in the realm of student-driven decision-making in target-oriented synthesis *(17)*.

4.2.8 Natural product extractions

It is possible for students to use a microwave oven to extract natural products from their biological sources. In 2017, the separation of three compounds (eugenol, eugenol acetate, and β-caryophyllene) from cloves was presented under conditions

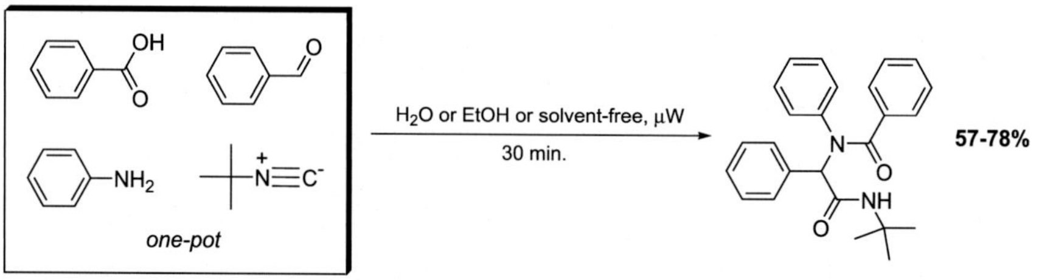

SCHEME 4.26

Ugi reaction forming a bisamide product.

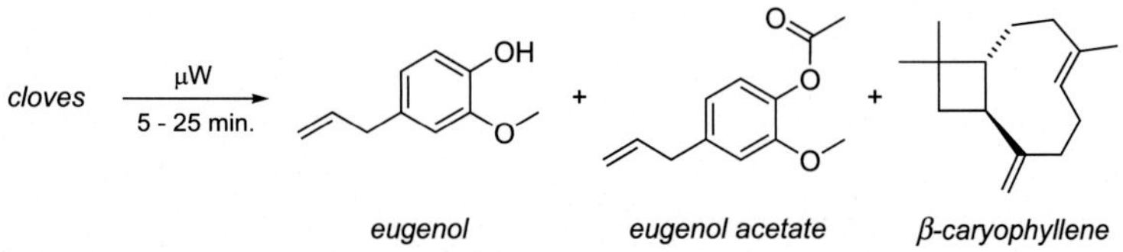

SCHEME 4.27

Extraction of three organic components from cloves.

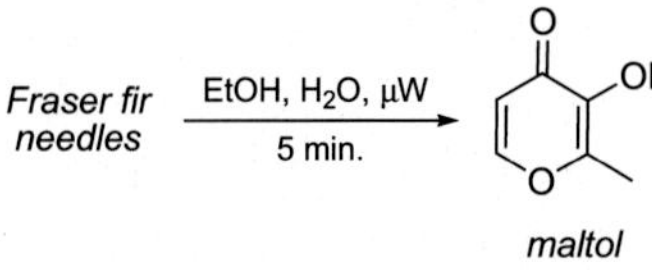

SCHEME 4.28

Extraction of maltol from Fraser fir needles.

of steam distillation, Soxhlet extraction, and microwave-assisted extraction *(72)*. The microwave heating time is varied between 5 and 25 min (Scheme 4.27), with an optimal separation of eugenol (47.5%) obtained on heating cloves at 70°C for 5 min in 100% ethanol. This compares with 49.1% of eugenol separated after 6 h via steam distillation, and only 3.7% separated after 4 h using a Soxhlet extractor (ethanol solvent). In a similar manner, Koch and coworkers used Fraser fir needles as a source of maltol by requiring students to compare and contrast a traditional extraction using dichloromethane with a microwave-assisted version using aqueous ethanol (Scheme 4.28). The authors note that "this experiment, more than any other, has fostered curiosity about the connections between organic chemistry, biology, and sustainability" *(73)*.

4.3 Considerations of reaction energy consumption in research and teaching laboratories

Somewhat surprisingly, relatively few quantitative studies have taken place in the research laboratory where the energy consumption of microwave reactions has been compared with that under conditions of conventional heating. In 2005, Gronnow et al. monitored the energy usage in preparing one mole of product from Suzuki couplings, a Friedel—Crafts acylation and a Knoevenagel condensation *(74)*, using a traditional oil bath and microwave reactors. The most significant finding was an 85-fold reduction in energy consumption on switching from an oil bath to a microwave oven for one of the Suzuki reactions. A similar trend was observed by Barnard and coworkers two years later, where continuous-flow microwave heating was found to be more energy-efficient for biodiesel production compared with more conventional apparatus *(75)*. In contrast, two groups noted the relative inefficiency for conversion of electrical to microwave energy (around 50%—65%), and results were published that indicated the scale on which reactions were performed was very significant, among other factors such as the type of reactor employed *(76,77)*. In some instances, microwave-heated reactions have been found to be significantly *less* energy efficient than their conventional counterparts. Indeed, a 2011 critical assessment of microwave-assisted organic synthesis concluded that "… the microwave heating process performed in laboratory-scale mono-mode microwave reactors is appallingly energy inefficient" *(78)*.

In addition to this research, Devine and Leadbeater probed conventional and microwave heating in batch mode for the synthesis of *N*-phenylpiperidine from aniline and 1,5-dibromopentane *(79)* and discovered the energy consumption to be comparable. However, at the meso scale ($\sim$1—3 L) it was determined that microwave heating could generally be more energy efficient than conventional heating for four pharmaceutically relevant organic reactions *(80)*. More recently, Cho et al. stated that "an automatic green label for microwave-assisted reactions is not warranted" after analyzing the energy consumption of heterogeneous catalytic reactions under conventional and microwave heating conditions *(81)*. The evident situation is summarized in a *Chemistry World* opinion piece written by Moseley, where he noted that assessing the energy of microwave-assisted reactions is complex and has to take into account many factors, to the point that each transformation must be considered on a case-by-case basis *(82)*.

In terms of education, the fact that energy consumptions are not routinely documented for literature protocols has been lamented *(83)*. Diehlmann et al. have established a set of "sustainable synthesis optimization" rules that are largely focused around energy considerations. These take into account factors such as thermal insulation of apparatus, reaction temperature/time, methods of energy input (e.g., conventional heating, microwaves or ultrasound), and specific heat capacities of solvents and auxiliaries *(84)*. An online and open-access database ("NOP") contains a number of laboratory reactions that can be used to teach principles of green chemistry to undergraduates. Several of these have had their energy efficiency determined by taking into account the total quantity of electrical energy consumed compared with the mass of the final product obtained *(85)*. However, of greatest relevance to the work presented in this chapter is the approach adopted by Stark et al. *(86)*. In this paper, an experiment is described for advanced undergraduates to investigate the synthesis of ionic liquids and to perform a detailed ecological assessment of different reaction conditions. As part of this, the energy demand of each reaction and workup is measured in kilojoules and expressed as an efficiency (taking into account the amount of ionic liquid synthesized (in kg/kJ)). From their measurements, students discover that contrasting a conductively heated synthesis with an identical reaction under microwave-assisted conditions indicates that the conventional method is more energetically favorable, despite the microwave reaction leading to a higher product yield.

4.4 Synopsis of CHM 343H ("Organic Synthesis Techniques"): An undergraduate course incorporating green chemistry thinking

At the University of Toronto, an undergraduate course for chemistry program students was designed in 2008 to showcase principles of modern organic synthesis, including those related to green chemistry and sustainability *(87)*. This offering runs from January to April each academic year (12 weeks in length) and includes 50 h of scheduled laboratory time. To supplement the practical work, a total of 24 classroom hours are included for discussion of experimental/spectroscopic methods, concepts pertaining to greener synthesis, and industrial case studies. The typical enrollment each spring has ranged from 30 to 40 undergraduates who are arranged in four laboratory sections (maximum of 12 students per section).

In order to avoid giving the impression to students that green chemistry is somehow a new subdiscipline or a compartmentalized area of chemical research, this course is simply titled "Organic Synthesis Techniques" (CHM 343H). A central pillar of CHM 343H is catalytic reactivity: students routinely undertake reactions that feature organocatalysis *(88)*, Brønsted and Lewis acid catalysis *(89,90)*, phase-transfer catalysis *(91)*, and transition-metal catalysis *(92–95)*. Their individual capacity to build green chemistry reasoning into a synthetic plan of their own design is determined via a capstone experiment toward the end of the course *(96)*, and a laboratory examination probes their ability to undertake a rudimentary life-cycle assessment *(97)*. CHM 343H is the centerpiece course in a unique University of Toronto undergraduate program of study entitled "Synthetic and Catalytic Chemistry" *(98)*.

One popular CHM 343H laboratory experiment has involved comparing two versions of a Biginelli reaction: a classical synthesis employing a solvent and a more modern, solvent-free approach *(90)*. Students assess the green features of the two methods by considering three of the 12 Principles of Green Chemistry *(16)*: atom economy, waste prevention, and design for energy efficiency (in a qualitative sense). In 2014, the Department of Chemistry purchased a commercial multimode microwave reactor for general research and teaching use *(99)* (Fig. 4.1). This prompted the Organic Synthesis Techniques teaching team to develop a robust, discovery-based experiment to quantify the energy consumption of a microwave-assisted reaction compared with an identical one undertaken via conventional reflux heating, in a similar vein to that reported by Stark et al. *(86)*. This was achieved through modification of an existing Suzuki synthesis to render it appropriate for a microwave reactor *(94)*.

4.5 CHM 343H microwave and conventional reflux experiment procedural overview

As mentioned in Section 4.2.6, the 2010 Chemistry Nobel Prize-winning Suzuki reaction has been incorporated into undergraduate curricula during the last

FIGURE 4.1

University of Toronto CEM MARS 6 multimode microwave oven for batch or parallel synthesis.

two decades *(66,67,94,100–113)*. The importance of Suzuki cross-couplings from an industrial perspective and notable reaction features (e.g., catalytic behavior and compatibility with benign solvents) has facilitated them as vehicles to teach principles of green chemistry. In 2008, Aktoudianakis and coworkers reported a facile preparation of 4-phenylphenol from phenylboronic acid and 4-iodophenol using water as the reaction solvent and palladium on carbon as the catalyst *(94)*. Under these conditions, the biaryl product is formed after 30 min of heating under reflux. This procedure was identified as being a potential candidate for microwave modification after consultation with faculty at the 2016 University of Oregon Green Chemistry Summer Workshop *(114)*. Most notably, the original reaction from 2008 was performed on a 1.0 mmol scale which was deemed to be too small for the MARS 6 Teflon reaction vessels and rotating carousel (Fig. 4.2). As such, the scale was doubled so that the reaction required 2.0 mmol of both 4-iodophenol and phenylboronic acid. This alteration meant that each CHM 343H student would be able to undertake a Suzuki reaction in duplicate via both microwave and conventional reflux heating during the same 4.5 h laboratory period. In the experimental design phase, reaction conditions were devised to ensure similar product yields via microwave irradiation (10 min) and reflux heating (30 min) (Scheme 4.29).

The experiment was introduced into the CHM 343H curriculum during the Spring 2017 academic semester and modified slightly for Spring 2018 when it was individually performed by a total of 40 undergraduates. Two laboratory sections (**#1** and **#2**, 10 students each) undertook practical activities on one day, with the remaining two sections (**#3** and **#4**, 10 students each) carrying out the identical experiment the following day. Students in sections **#1** and **#3** were instructed to

FIGURE 4.2

CEM MARS 6 microwave oven carousel and Teflon reaction vessel.

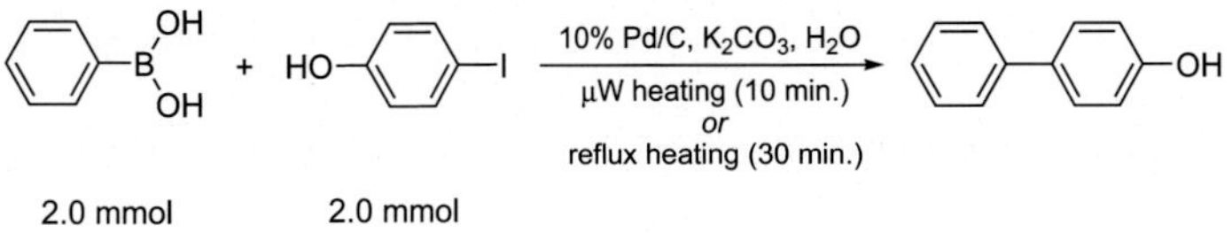

SCHEME 4.29

Suzuki synthesis of 4-phenylphenol under conditions of microwave and reflux heating.

initially set up their Suzuki reaction under conventional reflux conditions using a heating mantle. The total energy consumption of the mantle (in kWh) was directly measured for one reaction in each laboratory section using a Floureon power meter energy monitor *(115)*. During the 30-min reflux time, the same students each prepared their corresponding Suzuki microwave reaction in a Teflon tube. After working up their reflux reactions (isolating the biaryl products and purifying them by recrystallization), students placed their microwave tubes in the carousel (10 in total) which were heated to 130°C for 5 min ("ramp time") at an oven power of 800 W. Consequently, the temperature was held at 130°C for 10 min (the "fixed hold time"), and then cooling took place in the oven to room temperature (30 min). This latter time was used to discuss microwave heating theory with students and practical considerations and to explain the energy measurements being made. The total energy consumption for all 10 microwave reactions in one laboratory section (**#1** and **#3**) was directly recorded in kilowatt hours (1 kWh = 3.6 MJ) through an MTP 3100 wireless electricity monitor manufactured by MTP Inc. *(116)*. This particular monitor was chosen as it is able to directly measure the electrical power drawn by a microwave oven operating at 240 V. Students in sections **#2** and **#4** undertook exactly the same work but in reverse order, i.e., they started preparing the microwave reaction at the beginning of the practical session and finished with the conventional reflux reaction. Staggering of the oven usage in this manner was beneficial in facilitating small-group student discourse around microwave operations.

Each of the 40 students synthesized and purified the biaryl target compound under conditions of both conventional reflux and microwave irradiation. They calculated the percentage yields as a function of heating method and assessed product purity by measuring melting points and preparing samples for IR and proton NMR spectroscopy. Absolute energy measurements were provided by the instructor on a laboratory section basis with students having to compute energy consumption per mole of product. Following this, they wrote an extensive report which in part required them to engage with primary literature regarding the energy efficiency of microwave reactivity, along with their energy evaluations. More specifically, the following three postlaboratory questions were set:

1. One of the 12 Principles of Green Chemistry is "design for energy efficiency." Some energy measurements for the Suzuki reaction you have undertaken under both conventional reflux and microwave conditions are provided to you. Justifying your answer, comment on whether the synthesis of 4-phenylphenol is more energy efficient under microwave heating or via conventional heating under reflux.

2. Consult a research article (*ChemSusChem* **2008**, *1*, 123−132) and consider the microwave Suzuki reactions studied therein. Using the data presented in Table 4 of the article, and your answer to the first question, write a response to the statement that "microwave heating provides many advantages over conventional heating: one being an improvement in energy efficiency due to reduced reaction times."

3. Write a critical analysis (minimum of 250 words) that outlines the strengths and weaknesses of microwave chemistry, based solely on the Suzuki reactions that you have undertaken. What changes might you make to the Suzuki microwave procedure to improve its environmental profile?

4.6 Suzuki reaction results and discussion

Practical results obtained from CHM 343H students (average Suzuki product masses and reaction energy consumptions) are reported in Table 4.1. The biaryl product masses are calculated as averages across the entire class ($n = 40$) and correspond to yields of 54% (microwave irradiation) and 60% (conventional reflux). These values are consistent with those obtained during the development stages of this experiment by high school students and by a graduate student. Energy consumption for the reflux reaction using a heating mantle ranged from 0.042 to 0.056 kWh across the four laboratory sections (in good agreement with an average of 0.052 ± 0.006 kWh from experimental development work, $n = 4$). For microwave heating, the energy consumption varied from 1.222 to 1.353 kWh for each section

Table 4.1 Spring 2018 Suzuki reaction results summary.

Laboratory Section	Scale and heating method[a]	Reaction temperature (deg. C)	Reaction time (min.)[b]	Average product mass (mg)[c]	E (kWh)[d]	Energy consumed (kWh/mol)[e]
#1	μW	130	10	184	1.222	**113.0**
#1	Heating mantle	100	30	205	0.046	**38.2**
#2	μW	130	10	184	1.250	**115.6**
#2	Heating mantle	100	30	205	0.051	**42.3**
#3	μW	130	10	184	1.353	**125.1**
#3	Heating mantle	100	30	205	0.056	**46.5**
#4	μW	130	10	184	1.323	**122.4**
#4	Heating mantle	100	30	205	0.042	**34.9**

[a] All reactions undertaken at an identical scale (2.04 mmol 4-iodophenol, 2.05 mmol phenylboronic acid, 6.00 mmol K_2CO_3, 6 mg 10% Pd/C, 15 mL water).

[b] Reaction time defined as the time heating took place at the specified reaction temperature ("fixed hold time" for microwave heating).

[c] Average isolated product mass from CHM 343H student microwave reactions and conventional reflux reactions using a heating mantle across all demonstrator groups (40 students).

[d] For microwave heating: the total energy consumed (as measured by an electricity monitor) for **TEN** reactions per laboratory section. This includes the heating ramp time energy usage and the "cool-down" energy usage over 30 min at the end of the 10 min fixed hold time. For conventional reflux heating: the energy consumed for **ONE** heating mantle reaction as measured by a wattmeter. This includes the time required to heat the reaction to 100° C (reflux temperature).

[e] The energy consumed for **ONE** microwave reaction or **ONE** heating mantle reaction per mole of Suzuki product obtained, taking into account the average product mass in each case.

(10 reactions per section). This range compares favorably with 1.192 kWh and 1.297 kWh obtained during procedural testing.

In order to take into account the different average yields of the conventional reflux and microwave Suzuki reactions, the energy consumption was estimated by students on a per mole basis of biaryl product, following the approach of Razzaq and Kappe *(77)*. This afforded ranges of 34.9–46.5 kWh/mol for the reflux reactions (compared with 41.3 ± 6.1 kWh/mol in testing, $n = 7$) and 113.0–125.1 kWh/mol for the microwave reactions (compared with 107.3 ± 20.1 kWh/mol in testing, $n = 10$). To answer the first postlaboratory question, the CHM 343H students were largely able to calculate and interpret these values appropriately and to indicate that the synthesis of 4-phenylphenol was *more energy efficient* via *conventional heating under reflux compared with microwave heating on a per reaction basis.* From the values in Table 4.1, and those derived from experiment testing, the difference in energy consumption between the two heating methods can be approximated as a factor of 2.5–3.5.

The second postlaboratory question required students to access the same research article *(77)*, and to comment on experimental data regarding the energy efficiency of another Suzuki reaction (between phenylboronic acid and 4-bromoanisole under conditions of Pd catalysis). The authors of this publication investigated this transformation using several different heat sources: (1) a sealed vessel, single-mode microwave reactor; (2) an open vessel, single-mode microwave reactor; (3) an open vessel multimode microwave reactor; and (4) an isomantle. They additionally studied the reaction at different scales (1.5 mmol (microscale), 15 mmol (semimicroscale), and 150 mmol (macroscale)). At a microscale (similar to that undertaken by students themselves in the CHM 343H laboratory), the sealed vessel microwave irradiation process proved to be extremely efficient in terms of energy consumption (1.39 kWh/mol). However, at a semimicroscale, the total energy consumed was significantly lower by using an isomantle rather than an open vessel, single-mode microwave reactor (by approximately 57%), even with a longer heating time. Third, it was discovered that conventional heating was considerably more energy efficient than via an open vessel multimode microwave reactor on a macroscale. Taken together, these literature results clearly indicate it should not be assumed that microwave-assisted synthesis is automatically more energy efficient than processes that use more conventional heat sources. Student responses indicated that they were significantly able to draw this important conclusion based on their own laboratory data and that of Razzaq and Kappe *(77)*.

For the third and final question, students were tasked with drafting a critical summary of microwave reactivity, which was framed around their practical experience. They were additionally required to propose alterations to the microwave procedure they had followed in order to improve its environmental impact. A typical response to this question was as follows:

> *Compared to a conventional heating mantle, a microwave reaction chamber allows for many reactions to be run simultaneously in one batch. In the Suzuki*

cross-coupling experiment, ten reactions were run in each batch. However, the energy consumption for microwave synthesis (116 kWh/mol) is almost tripled versus the energy expense for reflux heating (42 kWh/mol). Since the microwave chamber can accommodate around 30 reaction tubes, energy efficiency can be improved by running more reactions in each batch. Furthermore, the microwave Suzuki reaction was complete within a reaction span that is one third of the time required for reflux heating. In pressurized microwave systems, polar solvents, such as water, can rapidly increase temperature far above their boiling points at atmospheric pressure. As a result of the greater energy input, a microwave reaction can reach completion faster than a reflux reaction that is conducted at 1 atm. A significantly reduced reaction time can be directly translated into energy efficiency, but it is certainly not the case for this experiment. Presumably, a notable amount of energy was consumed during the ramp-up and cooling stages during the microwave synthesis. In contrast, energy was not required to cool the reflux reaction mixture to room temperature. As a possible improvement in terms of energy efficiency, the reactions can be removed from the microwave and cooled at room temperature.

This particular student answer captures two important considerations regarding microwave technology. First, there is an energy cost associated with microwave cavity cooling to room temperature, as electricity is required to power a fan for the duration of the "cool-down" time (30 min in the case of the CHM 343H Suzuki reaction). Removing the carousel and microwave tubes from the reactor and allowing cooling to take place outside the oven would certainly reduce the overall energy consumption during the process. However, measurements made during the planning stages of this experiment indicate that only about one-third of the total microwave energy usage can be ascribed to the "cool-down," meaning that the conventional reflux reaction would still be more energy efficient. Second, in this experiment, the microwave carousel was not completely filled with sample tubes, as identified in the student response. Including more tubes should, in theory, improve the energy profile, but doing this does not take into account that in order to maintain the reaction parameters (in terms of time and product yield), the microwave power needs to be increased. Indeed, if more than 20 tubes are placed in the carousel, a power setting of 1800 W is required (compared with 800 W for 10 tubes). An interesting extension of this experiment would be to have students research the effect on overall energy consumption by varying the microwave power and the number of tubes in the reactor at a single time.

4.7 Conclusion

Many reactions that feature microwave technologies have appeared in the chemistry teaching literature during the last decade. The specific experiment described herein and run in course CHM 343H at the University of Toronto was designed to elicit student discussion around the relative energy efficiency of microwave chemistry. A persistent perception is that microwave heating is inherently more energy efficient than heating via conventional convection methods. Straightforward calculations and postlaboratory responses have indicated that students comprehend the problems associated with this notion, and that energy usage must be considered on a reaction-by-reaction basis. A contributing issue to this story is the relative inefficiency of converting electrical energy to microwave energy, a point that is perhaps underappreciated by educators and not communicated to undergraduates. The suggestion is made that curricular time be spent discussing the pros and cons of microwave transformations, rather than presenting them as watertight "green reactions." In short, let us not "make assumptions about energy consumptions."

4.8 Suzuki reaction student handout

4.8.1 Experiment objectives

1. To synthesize 4-phenylphenol from 4-iodophenol by an aqueous Suzuki reaction under both reflux and microwave heating conditions.
2. To learn about the potential advantages and potential disadvantages of microwave heating as a green chemistry approach, and to consider the energy efficiency of a Suzuki reaction under microwave heating compared to conventional reflux heating.

Today's Suzuki reaction combines phenylboronic acid with 4-iodophenol in the presence of Pd/C in order to synthesize a biaryl product under conditions of both microwave heating and conventional reflux (using a transformer and heating mantle). The two reactions will be undertaken at exactly the same scale and the energy consumption (in kilowatt hours (kWh)) will be measured for several reflux reactions and compared to that for the microwave reaction.

Compound	GMW	Amount added	mmol	mp (°C)	bp (°C)	d (g/mL)
Phenylboronic acid	121.93	250 mg	2.05	216–219		
4-Iodophenol	220.01	450 mg	2.04	92–94		
Potassium carbonate	138.21	830 mg	6.00	891		
10% Palladium on carbon	106.42	6 mg				
2M HCl	36.46	8 mL				
Methanol	32.04				65	0.792

4.8.2 Conventional reflux procedure

Place the following three solids into a 50-mL round-bottom flask equipped with a magnetic stir bar: phenylboronic acid (2.05 mmol), 4-iodophenol (2.04 mmol), and potassium carbonate (6.00 mmol). Add 10 mL of distilled water to the flask. Weigh 6 mg of Pd/C in a vial and rinse into the flask using 5 mL of distilled water. Reflux the mixture using a heating mantle and stir vigorously for 30 min **(set the transformer to 90 initially until the solvent boils, then turn down to 85 for the duration of the reflux period).** Add 2 M HCl (8 mL) while stirring with a glass rod to precipitate the reaction product. Collect the crude product (containing catalyst) on a Hirsch funnel by vacuum filtration and wash with a small amount of cold water (5 mL).

Completely dissolve the collected solid in 15 mL methanol (use a 25-mL Erlenmeyer flask) and remove the Pd/C by gravity filtration (collect the filtrate in a 50-mL Erlenmeyer flask). Add 10 mL of distilled water to the crude product dissolved in methanol, causing solid to precipitate. Heat until the entire product has gone into solution **(adding approximately 5 mL of water so that the total solvent volume is 30 mL and any cloudiness just disappears).** Once complete dissolution has occurred, allow the solution to cool slowly to room temperature and then cool in an ice bath. Collect the recrystallized product on a Hirsch funnel. Dry the product thoroughly, weigh and calculate the percentage yield, and obtain the product melting point. Prepare appropriate samples for IR spectroscopy (Nujol mull) and ^{1}H NMR spectroscopy (CDCl$_3$).

4.8.3 Microwave heating procedure

Place the following three solids into a 50-mL Erlenmeyer flask equipped with a magnetic stir bar: phenylboronic acid (2.05 mmol), 4-iodophenol (2.04 mmol), and potassium carbonate (6.00 mmol). Add 10 mL of distilled water to the flask and use 5 mL of additional distilled water to transfer 6 mg of palladium on carbon to the flask (preweigh this in a vial). Stir the reaction mixture rapidly for 15 min using a stir plate. Transfer the reaction mixture to a Teflon microwave tube. Attach the cap to the microwave tube and hand it to your demonstrator.

Each tube will be subjected to the following conditions in the microwave oven: (1) heat to 130°C for 5 min; (2) hold at 130°C for 10 min; (3) cool to room temperature (30 min). On recovering your tube, add 2M HCl (8 mL) while stirring with a glass rod. **Never use a metal spatula or a wire brush to stir/clean a Teflon microwave tube: use a glass rod instead!** Collect the precipitate by vacuum filtration using a Hirsch funnel and wash with a small amount of cold water (5 mL). Transfer the entire solid to a 25-mL Erlenmeyer flask and dissolve it in methanol (15 mL). Perform gravity filtration to remove the Pd/C catalyst and collect the filtrate in a 50-mL Erlenmeyer flask. Add 10 mL of distilled water to the crude product dissolved in methanol (causing solid to precipitate) and heat the solution to boiling. Heat until the entire product has gone into solution **(adding approximately 5 mL** of water so that the total solvent volume is 30 mL and any cloudiness just disappears). Remove the flask from the heat, and let it cool to room temperature and then in an ice bath. Collect the solid by vacuum filtration using a Hirsch funnel. Dry the recrystallized product thoroughly, weigh and calculate the percentage yield, and obtain the product melting point. Prepare appropriate samples for IR spectroscopy (Nujol mull) and ^1H NMR spectroscopy ($CDCl_3$).

Hand in the recrystallized product in a properly labeled glass vial. The label must include (1) **your full name**; (2) your demonstrator group number; (3) the structure (**not the name**) of the compound; (4) the mass of the material in the vial; and (5) the melting point of the material in the vial.

4.8.4 Hazards

1. Phenylboronic acid and 4-iodophenol are skin irritants and harmful if inhaled.
2. Potassium carbonate is irritating to the eyes, respiratory system, and skin.
3. Palladium on carbon is harmful if swallowed.
4. Methanol is highly flammable and toxic if swallowed.
5. Hydrochloric acid causes burns and is irritating to the respiratory system. Wear gloves, eye protection, and a laboratory coat at all times, and dispose of all chemicals in the appropriately labeled waste containers.

Acknowledgments

I wish to thank the following who made substantial contributions to the practical work described in this chapter: Victor Kobayashi, Timmy Seto, Alex Kerr, Jacob Salem, and Max Bennett (students of Michael Jansen at Crescent School, Toronto), and Dr. Sean Liew (University of Toronto). I am also grateful to the Department of Chemistry at the University of Toronto for a mini-Chemistry Teaching Fellowship to assist in laboratory development activities. Dr. Michael Koscho and Dr. Jared Mudrik are additionally acknowledged for their assistance at the 2016 University of Oregon Green Chemistry Summer Workshop and in Analest at the University of Toronto, respectively.

References

1. Ritter, S. K. Microwaves Stay Fast and Mystical. *Chem. Eng. News* **2014,** *92,* 26–28.
2. Leadbeater, N. E. Microwave-assisted Synthesis: General Concepts. In *Microwave-Assisted Polymer Synthesis;* Hoogenboom, R., Schubert, U. S., Wiesbrock, F., Eds.; Springer International Publishing: Cham, 2016; pp 1–44.
3. Kappe, C. O.; Pieber, B.; Dallinger, D. Microwave Effects in Organic Synthesis: Myth or Reality? *Angew. Chem. Int. Ed.* **2013,** *52,* 1088–1094.
4. Dudley, G. B.; Richert, R.; Stiegman, A. E. On the Existence of and Mechanism for Microwave-specific Reaction Rate Enhancement. *Chem. Sci.* **2015,** *6,* 2144–2152.

5. Prieto, P.; de la Hoz, A.; Diaz-Ortiz, A.; Rodriguez, A. M. Understanding MAOS Through Computational Chemistry. *Chem. Soc. Rev.* **2017,** *46,* 431−451.

6. Sharma, N.; Sharma, U. K.; Van der Eycken, E. V. Microwave-assisted Organic Synthesis: Overview of Recent Applications. In *Green Techniques for Organic Synthesis and Medicinal Chemistry;* Zhang, W., Cue, B. W., Eds., 2nd ed.; Wiley: Hoboken, NJ, 2018; pp 441−468.

7. de la Hoz, A.; Diaz-Ortiz, A.; Prieto, P. Microwave-assisted Green Organic Synthesis. In *Alternative Energy Sources for Green Chemistry;* Stefanidis, G., Stankiewicz, A., Eds.; Royal Society of Chemistry: Cambridge, 2016; pp 1−33.

8. Leadbeater, N. E. Organic Synthesis Using Microwave Heating. In *Comprehensive Organic Synthesis II;* Knochel, P., Ed., 2nd ed.; Elsevier: Amsterdam, 2014; pp 234−286.

9. Kokel, A.; Schäfer, C.; Török, B. Application of Microwave-assisted Heterogeneous Catalysis in Sustainable Synthesis Design. *Green Chem.* **2017,** *19,* 3729−3751.

10. Gawande, M. B.; Shelke, S. N.; Zboril, R.; Varma, R. S. Microwave-assisted Chemistry: Synthetic Applications for Rapid Assembly of Nanomaterials and Organics. *Acc. Chem. Res.* **2014,** *47,* 1338−1348.

11. Mehta, V. P.; Van der Eycken, E. V. Microwave-assisted C−C Bond Forming Cross-coupling Reactions: an Overview. *Chem. Soc. Rev.* **2011,** *40,* 4925−4936.

12. Polshettiwar, V.; Varma, R. S. Microwave-assisted Organic Synthesis and Transformations Using Benign Reaction Media. *Acc. Chem. Res.* **2008,** *41,* 629−639.

13. Dallinger, D.; Kappe, C. O. Microwave-assisted Synthesis in Water as Solvent. *Chem. Rev.* **2007,** *107,* 2563−2591.

14. Cresswell, S. L.; Haswell, S. J. Microwave Ovens−out of the Kitchen. *J. Chem. Educ.* **2001,** *78,* 900−904.

15. Zovinka, E. P.; Stock, A. E. Microwave Instruments: Green Machines for Green Chemistry? *J. Chem. Educ.* **2010,** *87,* 350−352.

16. Anastas, P. T.; Warner, J. C. *Green Chemistry: Theory and Practice;* Oxford University Press: New York, 1998; p 30.

17. Morra, B.; Dicks, A. P. Recent Progress in Green Undergraduate Organic Laboratory Design. In *Green Chemistry Experiments in Undergraduate Laboratories;* Fahey, J. T., Maelia, L. E., Eds.; American Chemical Society: Washington, DC, 2016; pp 7−32.

18. Dicks, A. P. A Review of Aqueous Organic Reactions for the Undergraduate Teaching Laboratory. *Green Chem. Lett. Rev.* **2009,** *2,* 9−21.

19. Dicks, A. P. Solvent-free Reactivity in the Undergraduate Organic Laboratory. *Green Chem. Lett. Rev.* **2009,** *2,* 87−100.

20. Moslet, S.; Shuster, D.; Cullen, M.; Henry, R. M. Interdisciplinary Bioinorganic Chemistry Laboratory: A Model Catecholate Siderophore With Fe(III). *Chem. Educ.* **2017,** *22,* 22−25.

21. Armstrong, C.; Burnham, J. A. J.; Warminski, E. E. Combining Sustainable Synthesis of a Versatile Ruthenium Dihydride Complex With Structure Determination Using Group Theory and Spectroscopy. *J. Chem. Educ.* **2017,** *94,* 928−931.

22. Crane, J. L.; Anderson, K. E.; Conway, S. G. Hydrothermal Synthesis and Characterization of a Metal−organic Framework by Thermogravimetric Analysis, Powder x-ray Diffraction, and Infrared Spectroscopy: An Integrative Inorganic Chemistry Experiment. *J. Chem. Educ.* **2015,** *92,* 373−377.

23. Nguyen, V. D.; Birdwhistell, K. R. Microwave Mapping Demonstration Using the Thermochromic Cobalt Chloride Equilibrium. *J. Chem. Educ.* **2014,** *91,* 880−882.

24. Dopke, J. A.; Dopke, N. C.; Kilpatrick, J. Microwave Synthesis of Chromium (III) Complexes With Acetylacetonate and Ethylenediamine Ligands. *Chem. Educ.* **2014,** *19,* 157−159.

25. Birdwhistell, K. R.; Conroy, K. J.; Schulz, B. E. Greening the Inorganic lab: Combining Microwaves and Phase Transfer Catalysis for the Rapid Synthesis of Group VI Carbonyl Complexes. *Chem. Educ.* **2014,** *19,* 133−137.

26. Arnold, A. M.; Kwak, D. J.; Lofgren, L. E.; Walters, B. M.; Wilt, A. L.; Woldemeskel, S. A.; Zovinka, E. P. Microwaving Metals: Inserting Metals Into Porphyrin Ligands Using Microwave Methods. *Chem. Educ.* **2014,** *19,* 299−301.

27. Leyral, G.; Bernaud, L.; Manteghetti, A.; Filhol, J.-S. Microwave Synthesis of a Fluorescent Ruby Powder. *J. Chem. Educ.* **2013,** *90,* 1380−1383.

28. Foreman, D. J.; Horta, J. E. Determination of the Empirical Formula of Zinc Bromide Using Microwave Technology: A Simple Experiment for the Undergraduate General Chemistry Laboratory. *Chem. Educ.* **2013,** *18,* 302−303.

29. Sharma, R. K.; Sharma, C.; Sidhwani, I. T. Solventless and One-pot Synthesis of Cu(II) Phthalocyanine Complex: A Green Chemistry Experiment. *J. Chem. Educ.* **2011,** *88,* 86−87.

30. Zitoun, D.; Bernaud, L.; Manteghetti, A.; Filhol, J.-S. Microwave Synthesis of a Long-lasting Phosphor. *J. Chem. Educ.* **2009,** *86,* 72−75.

31. Berry, D. E.; Fawkes, K. L. Solid State Isomerization. *Chem. Educ.* **2008,** *13,* 158−160.

32. Yoshikawa, N.; Takashima, H. Microwave-assisted Dehydration of Aqua Complexes. *Chem. Educ.* **2002,** *7,* 354−355.

33. Ardon, M.; Hayes, P. D.; Hogarth, G. Microwave-assisted Reflux in Organometallic Chemistry: Synthesis and Structural Determination of Molybdenum Carbonyl Complexes. An Intermediate-level Organometallic-inorganic Experiment. *J. Chem. Educ.* **2002,** *79,* 1249−1251.

34. Bowden, J. A.; Nocito, B. A.; Lowers, R. H.; Guillette, L. J.; Williams, K. R.; Young, V. Y. Environmental Indicators of Metal Pollution and Emission: An Experiment for the Instrumental Analysis Laboratory. *J. Chem. Educ.* **2012,** *89,* 1057−1060.

35. Schaber, P. M.; Pines, H. A.; Larkin, J. E.; Shepherd, L. A.; Wierchowski, E. E. The Case of nut Poisoning (or Too Much of a Good Thing?): Implementation and Assessment. *J. Chem. Educ.* **2011,** *88,* 1012−1013.

36. Nahir, T. M.; Sheffield, M.-C. Analysis of Selenium in Brazil Nuts by Microwave Digestion and Fluorescence Detection. *J. Chem. Educ.* **2002,** *79,* 1345−1347.

37. Pham, S. N.; Kuether, J. E.; Gallagher, M. J.; Hernandez, R. T.; Williams, D. N.; Zhi, B.; Mensch, A. C.; Hamers, R. J.; Rosenzweig, Z.; Fairbrother, H.; Krause, M. O. P.; Feng, Z. V.; Haynes, C. L. Carbon Dots: A Modular Activity to Teach Fluorescence and Nanotechnology at Multiple Levels. *J. Chem. Educ.* **2017,** *94,* 1143−1149.

38. Paniconi, G.; Chung, S.-J.; Liou, S.-C.; Kim, S. S.; Kim, L.; Kang, S.; Kim, D. Three Green Approaches to the Synthesis of Silver Nanoparticles Using Pentose and Hexose Carbohydrates. *Chem. Educ.* **2016,** *21,* 73−76.

39. Cooke, J.; Hebert, D.; Kelly, J. A. Sweet Nanochemistry: A Fast, Reliable Alternative Synthesis of Yellow Colloidal Silver Nanoparticles Using Benign Reagents. *J. Chem. Educ.* **2015,** *92,* 345−349.

40. Tian, J.; Yan, L.; Sang, A.; Yuan, H.; Zheng, B.; Xiao, D. Microwave-assisted Synthesis of Red-light Emitting Au Nanoclusters With the Use of Egg White. *J. Chem. Educ.* **2014,** *91,* 1715−1719.
41. Dziedzic, R. M.; Gillian-Daniel, A. L.; Petersen, G. M.; Martínez-Hernández, K. J. Microwave Synthesis of Zinc Hydroxy Sulfate Nanoplates and Zinc Oxide Nanorods in the Classroom. *J. Chem. Educ.* **2014,** *91,* 1710−1714.
42. Dong, Z.; Richardson, D.; Pelham, C.; Islam, M. R. Rapid Synthesis of Silver Nanoparticles Using a Household Microwave and Their Characterization: a Simple Experiment for Nanoscience. *Chem. Educ.* **2008,** *13,* 240−243.
43. Baar, M. R. Greener Organic Reactions Under Microwave Heating. In *Green Organic Chemistry in Lecture and Laboratory;* Dicks, A. P., Ed.; CRC Press: Boca Raton, FL, 2012; pp 225−256.
44. *J. Chem. Educ.* **2018.** Author Guidelines. http://pubs.acs.org/paragonplus/submission/jceda8/jceda8_authguide.pdf.
45. Russell, C. B.; Mason, J. D.; Bean, T. G.; Murphree, S. S. A Student-centered First-semester Introductory Organic Laboratory Curriculum Facilitated by Microwave-assisted Synthesis (MAOS). *J. Chem. Educ.* **2014,** *91,* 511−517.
46. Mendoza, M. S.; Sanchez, A. V.; Manrique, C. G.; Avila-Zarraga, J. G. Reduction of Nitro Compounds Using the Pd/H_2N-NH_2/Microwave System. *Educ. Quím.* **2013,** *24,* 347−350.
47. Pilcher, S. C.; Coats, J. Preparing 4-ethoxyphenylurea Using Microwave Irradiation: Introducing Students to the Importance of Artificial Sweeteners and Microwave-assisted Organic Synthesis (MAOS). *J. Chem. Educ.* **2017,** *94,* 260−263.
48. Anderluh, M. Synthesis of Celecoxib in the Undergraduate Organic Chemistry or Medicinal Chemistry Laboratories; Conventional vs. Microwave Heating. *Chem. Educ.* **2013,** *18,* 155−158.
49. Keuseman, K. J.; Morrow, N. C. A "Green" Approach to Synthesis of Trans-4-methoxycinnamic Acid in the Undergraduate Teaching Laboratory. *Chem. Educ.* **2014,** *19,* 347−350.
50. Baar, M. R. Research Experience for the Organic Chemistry Laboratory: A Student-centered Optimization of a Microwave-enhanced Williamson Ether Synthesis and GC Analysis. *J. Chem. Educ.* **2018,** *95,* 1235−1237.
51. Baar, M. R.; Gammerdinger, W.; Leap, J.; Morales, E.; Shikora, J.; Weber, M. H. Pedagogical Comparison of Five Reactions Performed Under Microwave Heating in Multi-mode Versus Mono-mode Ovens: Diels−Alder Cycloaddition, Wittig Salt Formation, E2 Dehydrohalogenation to Form an Alkyne, Williamson Ether Synthesis, and Fischer Esterification. *J. Chem. Educ.* **2014,** *91,* 1720−1724.
52. Jensen, J.; Grundy, S. C.; Bretz, S. L.; Hartley, C. S. Synthesis and Characterization of Self-assembled Liquid Crystals: p-alkoxybenzoic Acids. *J. Chem. Educ.* **2011,** *88,* 1133−1136.
53. Hu, J.; Yin, J.; Lin, T.; Li, G. Synthesis of an Ionic Liquid and Its Application as Template for the Preparation of Mesoporous Material MCM-41: a Comprehensive Experiment for Undergraduate Students. *J. Chem. Educ.* **2012,** *89,* 284−285.
54. Lampkowski, J. S.; Bass, W.; Nimmo, Z.; Young, D. D. Microwave-assisted Esterifications: An Unknowns Experiment Designed for an Undergraduate Organic Chemistry Laboratory. *World J. Chem. Educ.* **2015,** *3,* 111−114.

55. Reilly, M. K.; King, R. P.; Wagner, A. J.; King, S. M. Microwave-assisted Esterification: A Discovery-based Microscale Laboratory Experiment. *J. Chem. Educ.* **2014,** *91,* 1706−1709.

56. Fahey, J. T.; Dineen, A. E.; Henain, J. M. Microwave-assisted Aspirin Synthesis From Over-the-counter Pain Creams Using Naturally Acidic Catalysts: A Green Undergraduate Organic Chemistry Laboratory Experiment. In *Green Chemistry Experiments in Undergraduate Laboratories,* Vol. 1233, Fahey, J. T., Maelia, L. E., Eds.; American Chemical Society: Washington, DC, 2016; pp 93−109.

57. Shell, T. A.; Shell, J. R.; Poole, K. A.; Guetzloff, T. F. Microwave-assisted Synthesis of N-phenylsuccinimide. *J. Chem. Educ.* **2011,** *88,* 1439−1441.

58. Horta, J. E. Simple Microwave-assisted Claisen and Dieckmann Condensation Experiments for the Undergraduate Organic Chemistry Laboratory. *J. Chem. Educ.* **2011,** *88,* 1014−1015.

59. Priya, T. J.; Sarkar, S.; Shamili, G.; Sugumar, R. W. Microwave Assisted Depolymerization of Post Consumer PET Waste Using Phase Transfer Catalysts. *Chem. Educ.* **2014,** *19,* 61−63.

60. Pohl, N. L. B.; Streff, J. M.; Brokman, S. Evaluating Sustainability: Soap Versus Biodiesel Production From Plant Oils. *J. Chem. Educ.* **2012,** *89,* 1053−1056.

61. Latimer, D.; Wiebe, M. Greening the Organic Chemistry Laboratory: A Comparison of Microwave-assisted and Classical Nucleophilic Aromatic Substitution Reactions. *Green Chem. Lett. Rev.* **2015,** *8,* 39−42.

62. Damkaci, F.; Dallas, M.; Wagner, M. A Microwave-assisted Friedel−Crafts Acylation of Toluene With Anhydrides. *J. Chem. Educ.* **2013,** *90,* 390−392.

63. Yadav, U.; Mande, H.; Ghalsasi, P. Nitration of Phenols Using $Cu(NO_3)_2$: Green Chemistry Laboratory Experiment. *J. Chem. Educ.* **2012,** *89,* 268−270.

64. Saunders, P. R. C.; Al-Jobory, Y. M.; Hwangbo, J.; Schultz, S. J.; Merbouh, N. Rapid Microwave-assisted Isomerization of cis-norbornene-endo-5,6-dicarboxylic Anhydride. *Chem. Educ.* **2011,** *16,* 149−151.

65. DeFrancesco, H.; Dudley, J.; Coca, A. Undergraduate Organic Experiment: Tetrazole Formation by Microwave Heated (3 + 2) Cycloaddition in Aqueous Solution. *J. Chem. Educ.* **2018,** *95,* 433−437.

66. Vargas, B. P.; Rosa, C. H.; da Silva Rosa, D.; Rosa, G. R. "Green" Suzuki-Miyaura Cross-coupling: an Exciting Mini-project for Chemistry Undergraduate Students. *Educ. Quím.* **2016,** *27,* 139−142.

67. Soares, P.; Fernandes, C.; Chavarria, D.; Borges, F. Microwave-assisted Synthesis of 5-phenyl-2-hydroxyacetophenone Derivatives by a Green Suzuki Coupling Reaction. *J. Chem. Educ.* **2015,** *92,* 575−578.

68. Ingold, M.; Colella, L.; Dapueto, R.; Lopez, G. V.; Porcal, W. Ugi Four-component Reaction (U-4CR) Under Green Conditions Designed for Undergraduate Organic Chemistry Laboratories. *World J. Chem. Educ.* **2017,** *5,* 153−157.

69. McElroy, C. R.; Constantinou, A.; Jones, L. C.; Summerton, L.; Clark, J. H. Towards a Holistic Approach to Metrics for the 21st Century Pharmaceutical Industry. *Green Chem.* **2015,** *17,* 3111−3121.

70. Pohl, N. L. B.; Kirshenbaum, K.; Yoo, B.; Schulz, N.; Zea, C. J.; Streff, J. M.; Schwarz, K. L. Student-driven Design of Peptide Mimetics: Microwave-assisted Synthesis of Peptoid Oligomers. *J. Chem. Educ.* **2011,** *88,* 999−1001.

71. Slade, M. C.; Raker, J. R.; Kobilka, B.; Pohl, N. L. B. A Research Module for the Organic Chemistry Laboratory: Multistep Synthesis of a Fluorous dye Molecule. *J. Chem. Educ.* **2014,** *91,* 126−130.
72. Guntero, V. A.; Mancini, P. M.; Kneeteman, M. N. Introducing Organic Chemistry Students to the Extraction of Natural Products Found in Vegetal Species. *World J. Chem. Educ.* **2017,** *5,* 142−147.
73. Koch, A. S.; Chimento, C. A.; Berg, A. N.; Mughal, F. D.; Spencer, J.-P.; Hovland, D. E.; Mbadugha, B.; Hovland, A. K.; Eller, L. R. Extraction of Maltol From Fraser fir: A Comparison of Microwave-assisted Extraction and Conventional Heating Protocols for the Organic Chemistry Laboratory. *J. Chem. Educ.* **2015,** *92,* 170−174.
74. Gronnow, M. J.; White, R. J.; Clark, J. H.; Macquarrie, D. J. Energy Efficiency in Chemical Reactions: A Comparative Study of Different Reaction Techniques. *Org. Process Res. Dev.* **2005,** *9,* 516−518.
75. Barnard, T. M.; Leadbeater, N. E.; Boucher, M. B.; Stencel, L. M.; Wilhite, B. A. Continuous-flow Preparation of Biodiesel Using Microwave Heating. *Energy Fuels* **2007,** *21,* 1777−1781.
76. Hoogenboom, R.; Wilms, T. F. A.; Erdmenger, T.; Schubert, U. S. Microwave-assisted Chemistry: A Closer Look at Heating Efficiency. *Aust. J. Chem.* **2009,** *62,* 236−243.
77. Razzaq, T.; Kappe, C. O. On the Energy Efficiency of Microwave-assisted Organic Reactions. *ChemSusChem* **2008,** *1,* 123−132.
78. Moseley, J. D.; Kappe, C. O. A Critical Assessment of the Greenness and Energy Efficiency of Microwave-assisted Organic Synthesis. *Green Chem.* **2011,** *13,* 794−806.
79. Devine, W. G.; Leadbeater, N. E. Probing the Energy Efficiency of Microwave Heating and Continuous-flow Conventional Heating as Tools for Organic Chemistry. *ARKIVOC* **2011,** 127−143.
80. Moseley, J. D.; Woodman, E. K. Energy Efficiency of Microwave- and Conventionally Heated Reactors Compared at Meso Scale for Organic Reactions. *Energy Fuels* **2009,** *23,* 5438−5447.
81. Cho, H.; Török, F.; Török, B. Energy Efficiency of Heterogeneous Catalytic Microwave-assisted Organic Reactions. *Green Chem.* **2014,** *16,* 3623−3634.
82. Moseley, J. D. Microwave Chemistry − Green or not? *Chem. World* **February 24, 2011,** *8* (3), 34.
83. Andraos, J.; Dicks, A. P. Green Chemistry Teaching in Higher Education: AReview of Effective Practices. *Chem. Educ. Res. Pract.* **2012,** *13,* 69−79.
84. Diehlmann, A.; Kreisel, G.; Gorges, R. Contribution to "Developing Sustainability" in Chemical Education. *Chem. Educ.* **2003,** *8,* 102−106.
85. Eissen, M.; Bahadir, M.; König, B.; Ranke, J. Developing and Disseminating NOP: An Online, Open-access, Organic Chemistry Teaching Resource to Integrate Sustainability Concepts in the Laboratory. *J. Chem. Educ.* **2008,** *85,* 1000−1005.
86. Stark, A.; Ott, D.; Kralisch, D.; Kreisel, G.; Ondruschka, B. Ionic Liquids and Green Chemistry: A lab Experiment. *J. Chem. Educ.* **2010,** *87,* 196−201.
87. Dicks, A. P.; Batey, R. A. ConfChem Conference on Educating the Next Generation: Green and Sustainable Chemistry−greening the Organic Curriculum: Development of an Undergraduate Catalytic Chemistry Course. *J. Chem. Educ.* **2013,** *90,* 519−520.
88. Stabile, R. G.; Dicks, A. P. Two-step Semi-microscale Preparation of a Cinnamate Ester Sunscreen Analog. *J. Chem. Educ.* **2004,** *81,* 1488−1491.

89. Gómez-Biagi, R. F.; Dicks, A. P. Assessing Process Mass Intensity and Waste via an aza-Baylis-Hillman Reaction. *J. Chem. Educ.* **2015,** *92,* 1938−1942.

90. Aktoudianakis, E.; Chan, E.; Edward, A. R.; Jarosz, I.; Lee, V.; Mui, L.; Thatipamala, S. S.; Dicks, A. P. Comparing the Traditional With the Modern: A Greener, Solvent-free Dihydropyrimidone Synthesis. *J. Chem. Educ.* **2009,** *86,* 730−732.

91. Goodreid, J. D.; Dicks, A. P. SpinWorks Software as an Organic Undergraduate NMR Processing Tool. *Chem. Educ.* **2012,** *17,* 133−136.

92. Koroluk, K. J.; Skonieczny, S.; Dicks, A. P. Microscale Catalytic and Chemoselective TPAP Oxidation of Geraniol. *Chem. Educ.* **2011,** *16,* 307−309.

93. Obhi, N. K.; Mallov, I.; Borduas-Dedekind, N.; Rousseaux, S. A. L.; Dicks, A. P. Comparing Industrial Amination Reactions in a Combined Class and Laboratory Green Chemistry Assignment. *J. Chem. Educ.* **2019,** *96,* 93−99.

94. Aktoudianakis, E.; Chan, E.; Edward, A. R.; Jarosz, I.; Lee, V.; Mui, L.; Thatipamala, S. S.; Dicks, A. P. "Greening up" the Suzuki Reaction. *J. Chem. Educ.* **2008,** *85,* 555−557.

95. Cheung, L. L. W.; Aktoudianakis, E.; Chan, E.; Edward, A. R.; Jarosz, I.; Lee, V.; Mui, L.; Thatipamala, S. S.; Dicks, A. P. A Microscale Heck Reaction in Water. *Chem. Educ.* **2007,** *12,* 77−79.

96. Edgar, L. J. G.; Koroluk, K. J.; Golmakani, M.; Dicks, A. P. Green Chemistry Decision-making in an Upper-level Undergraduate Organic Laboratory. *J. Chem. Educ.* **2014,** *91,* 1040−1043.

97. Dicks, A. P.; Hent, A.; Koroluk, K. J. The EcoScale as a Framework for Undergraduate Green Chemistry Teaching and Assessment. *Green Chem. Lett. Rev.* **2018,** *11,* 29−35.

98. *University of Toronto Synthetic and Catalytic Chemistry Undergraduate Program.* www.chem.utoronto.ca/undergrad/syntheticcatalytic.php.

99. *CEM MARS™ 6 Multi-mode Microwave Oven for Batch or Parallel Synthesis.* http://cem.com/en/mars-6.

100. Pullen, R.; Olding, A.; Smith, J. A.; Bissember, A. C. Capstone Laboratory Experiment Investigating Key Features of Palladium-catalyzed Suzuki−Miyaura Cross-coupling Reactions. *J. Chem. Educ.* **2018,** *95,* 2081−2085.

101. Thananatthanachon, T.; Lecklider, M. R. Synthesis of Dichlorophosphinenickel(II) Compounds and Their Catalytic Activity in Suzuki Cross-coupling Reactions: A Simple air-free Experiment for Inorganic Chemistry Laboratory. *J. Chem. Educ.* **2017,** *94,* 786−789.

102. Christensen, D.; Cohn, P. G. Minding the gap: Synthetic Strategies for Tuning the Energy gap in Conjugated Molecules. *J. Chem. Educ.* **2016,** *93,* 1794−1797.

103. Oliveira, D. G. M.; Rosa, C. H.; Vargas, B. P.; Rosa, D. S.; Silveira, M. V.; de Moura, N. F.; Rosa, G. R. Introducing Undergraduates to Research Using a Suzuki−Miyaura Cross-coupling Organic Chemistry Miniproject. *J. Chem. Educ.* **2015,** *92,* 1217−1220.

104. Hie, L.; Chang, J. J.; Garg, N. K. Nickel-catalyzed Suzuki−Miyaura Cross-coupling in a Green Alcohol Solvent for an Undergraduate Organic Chemistry Laboratory. *J. Chem. Educ.* **2015,** *92,* 571−574.

105. Hill, N. J.; Bowman, M. D.; Esselman, B. J.; Byron, S. D.; Kreitinger, J.; Leadbeater, N. E. Ligand-free Suzuki−Miyaura Coupling Reactions Using an Inexpensive Aqueous Palladium Source: A Synthetic and Computational Exercise for the Undergraduate Organic Chemistry Laboratory. *J. Chem. Educ.* **2014,** *91,* 1054−1057.

106. Heiskanen, J. P. Synthesis of Organic Semiconductor Backbone Using a Suzuki-Miyaura Cross-coupling. *Chem. Educ.* **2014,** *19,* 275−277.

107. Hamilton, A. E.; Buxton, A. M.; Peeples, C. J.; Chalker, J. M. An Operationally Simple Aqueous Suzuki−Miyaura Cross-coupling Reaction for an Undergraduate Organic Chemistry Laboratory. *J. Chem. Educ.* **2013,** *90,* 1509−1513.

108. Costa, N. E.; Pelotte, A. L.; Simard, J. M.; Syvinski, C. A.; Deveau, A. M. Discovering Green, Aqueous Suzuki Coupling Reactions: Synthesis of Ethyl (4-phenylphenyl)acetate, a Biaryl With Anti-arthritic Potential. *J. Chem. Educ.* **2012,** *89,* 1064−1067.

109. Pantess, D. A.; Rich, C. V. Aqueous Suzuki Reactions: a Greener Approach to Transition Metal-mediated Aryl Couplings in the Organic Instructional Laboratory. *Chem. Educ.* **2009,** *14,* 258−260.

110. Novak, M.; Wang, Y.-T.; Ambrogio, M. W.; Chan, C. A.; Davis, H. E.; Goodwin, K. S.; Hadley, M. A.; Hall, C. M.; Herrick, A. M.; Ivanov, A. S.; Mueller, C. M.; Oh, J. J.; Soukup, R. J.; Sullivan, T. J.; Todd, A. M. A Research Project in the Organic Instructional Laboratory Involving the Suzuki-Miyaura Cross Coupling Reaction. *Chem. Educ.* **2007,** *12,* 414−418.

111. Grove, T.; DiLella, D.; Volker, E. Stereospecific Synthesis of the Geometrical Isomers of a Natural Product. *J. Chem. Educ.* **2006,** *83,* 1055−1057.

112. Hoogenboom, R.; Meier, M. A. R.; Schubert, U. S. The Introduction of High-throughput Experimentation Methods for Suzuki−Miyaura Coupling Reactions in University Education. *J. Chem. Educ.* **2005,** *82,* 1693−1696.

113. Callam, C. S.; Lowary, T. L. Suzuki Cross-coupling Reactions: Synthesis of Unsymmetrical Biaryls in the Organic Laboratory. *J. Chem. Educ.* **2001,** *78,* 947−948.

114. *2016 University of Oregon Green Chemistry Summer Workshop.* https://communities.acs.org/events/2146.

115. *Floureon Power Meter Energy Monitor.* www.floureon.com/product-g_142.html.

116. *MTP Inc. Wireless Electricity Monitor (MTP 3100).* www.mtpinc.com/Files/MTP_3100_Gen_2_Manual_2018_EN.pdf.

Making connections: Implementing a community-based learning experience in green chemistry

Renuka Manchanayakage, PhD

Assistant Professor, Department of Chemistry, St. John Fisher College, Rochester, NY, United States

5.1 Introduction

Experiential learning has been identified as an important pedagogical method *(1)*. Among many types of experiential learning, much attention has been given to community-based learning or service-learning. According to the definition given by Jacoby, community-based service-learning is "a form of experiential education in which students engage in activities that address human and community needs together with structured opportunities intentionally designed to promote students learning and development *(2)*." The categories of community-based service-learning include fieldwork, fellowships, internships, volunteering, and clinical experiences. This type of pedagogy provides students with the opportunity to apply the knowledge and skills that they learn in the classroom to solve a real-world issue and is considered a high-impact pedagogical practice *(3)*. When working with a community partner, students can sometimes be exposed to different career options. A meaningful community experience can also help students develop civic responsibility and foster a sense of caring for others *(4)*.

5.1.1 Community-based service-learning in chemistry

Incorporation of community-based service-learning into chemistry courses has been reported by many universities *(5−7)*. Extra time, effort, and resources required for adding a service-learning component are often offset by the depth of student learning and service to the community. Service-learning can supplement the chemistry education that students receive during lectures and help to develop a lasting appreciation for the subject matter. The majority of these service-learning activities involve environmental soil, water, and air quality monitoring. In addition, the scientific education of younger generations using different activities and workshops has been demonstrated.

Integrating Green and Sustainable Chemistry Principles into Education. https://doi.org/10.1016/B978-0-12-817418-0.00005-X
Copyright © 2019 Elsevier Inc. All rights reserved.

Several institutions have developed service-learning chemistry courses in which undergraduate students have taught chemistry experiments or performed scientific demonstrations for secondary school students *(8–10)*. As an example, the general chemistry students at the Claremont Colleges developed a chemical demonstration series and presented it at the Chemistry Day at Serrano Middle School. The demonstrations were performed in the outdoor courtyard in a carnival atmosphere before the sixth, seventh, and eighth grade students. Through this activity, college students were able to utilize the knowledge they gained and middle school students were able to learn chemistry and about life in the college *(10)*.

A number of reports have outlined the incorporation of environmental analyses as a community-based learning experience *(11–14)*. An analytical chemistry laboratory course offered at the University of Central Florida has been modified to include a community learning experience based on water quality analysis *(11)*. Students enrolled in this course have studied the wastewater effluent at the Orlando Easterly Wetlands which is an engineered water polishing facility designed to remove nutrients from treated wastewater. Students have sampled water and performed analysis on it, including measurement of pH, chloride levels, total dissolved solids, phosphorus levels, and amounts of sucralose (an artificial sweetener). Students have also shared their results in a public seminar. At the end of the course, the student learning of chemistry content has been assessed by a final examination. This has been administered to both the service-learning section and the conventional section of the class, with the average grade being higher for the service-learning section *(11)*.

In addition to environmental analysis and educating younger generations, other community-based learning activities have been integrated into organic and biochemistry courses *(15–17)*. At Central Washington University, a study compared learning outcomes of a traditional Introduction to Biochemistry course to one that uses a community-based inquiry instructional method. The community-based inquiry curriculum included laboratory investigations on local elementary school lunch and human health case studies. Students from the community-based inquiry section demonstrated high content knowledge and significant critical thinking compared with students from the traditional course *(17)*. All evidence supports the notion that community-based service-learning provides many benefits including increased academic and cognitive skills, higher ability to apply course content, good leadership qualities, and increased community engagement.

5.2 Green chemistry education

Green chemistry has evolved to become a mainstream practice in academia and industry *(18,19)*. Green chemistry is the design of chemical products and processes that reduce or eliminate the use and formation of hazardous chemicals *(20)*. Therefore, green chemistry practices are intended to provide economic and environmental benefits to society as a whole. With growing public concern about environmental issues, current students are extremely interested in learning about green chemistry and sustainability concepts *(21)*. As a result, many universities have integrated green chemistry principles into courses or research endeavors *(22–24)*. For example, efficient approaches of incorporating green chemistry concepts into existing general chemistry and organic chemistry courses have been demonstrated *(25–27)*. The cross-disciplinary nature of green chemistry also makes it possible to be taught in conjunction with other subjects, and many new green chemistry courses have been developed and integrated into curricula. Some of these courses have been designed for chemistry or science majors, while others have targeted nonscience majors *(28–30)*. A community-based service-learning project has been introduced into an undergraduate green chemistry course offered at Westminster College *(31)*. In this project, students have developed green chemistry laboratory activities to be used in a local high school. Even though green chemistry concepts are significantly related to environmental and societal aspects, there are few documented attempts of integrating community-based service-learning into green chemistry courses.

5.3 St. John Fisher College

A green chemistry course including a community-based learning experience is offered at St. John Fisher College (SJFC) within undergraduate sustainability programs. SJFC was founded in 1948 and is an independent, liberal arts institution located in Rochester, NY *(32)*. SJFC is committed to reducing the carbon footprint of the college and opened its first LEED-certified educational facility (the Integrated Science and Health Sciences Building) in Fall 2015. SJFC also established the Center for Sustainability with the mission of advancing sustainable practices across campus and throughout the community to generate positive economic, social, and environmental outcomes. The college offers an interdisciplinary sustainability major program and a minor program. These programs are designed to explore the theory and practice of sustainability through the lens of several different disciplines. Additionally, the Center for Service-Learning at SJFC oversees the community-based service-learning activities at the college. This center provides training for faculty who plan to integrate service-learning into their courses. It also works as a liaison between faculty and community partners to find the right community partner for each course.

5.3.1 Green chemistry education at SJFC

The developed green chemistry course is offered as an elective under both sustainability major and minor programs. This is a freshman-level course which can also be used to fulfill a general science requirement for nonscience majors who have to complete a three-credit science course. Therefore, the course is usually composed of a diverse group of students including science and nonscience majors. While some students have taken chemistry or other science courses, some have never taken any college-level science course. This makes teaching the course both interesting and challenging. The course covers sustainability from a chemistry perspective and is designed to instill an appreciation for green chemistry and sustainability within a diverse group of students who may be future scientists, business leaders, or policymakers.

In 2017, the green chemistry course was redesigned by integrating a community-based service-learning component. Green chemistry inherently has environmental and societal aspects, and as such community-based service-learning is a natural fit. Students can develop civic pride and a sense of responsibility by being involved and invested to help the community. The course is offered in a workshop style by combining lectures with service-learning and/or laboratory activities. The classes are offered on Tuesdays and Thursdays and are each 90 min in length. There are no separate laboratory sections, and any practical or field activities related to service-learning projects are completed during the 90-minute class periods. If any project requires more than 90 min to complete, they are divided among two class periods. There are also a few occasions when some students stay after the class period to complete particular tasks.

The class schedule consists of five major units of green chemistry: (1) introduction to green chemistry; (2) air; (3) water; (4) energy; and (5) polymers and plastics. Lectures are based on two textbooks: (1) *Chemistry in Context: Applying Chemistry to Society* (8th Ed.) by the American Chemical Society and (2) *Energy: Its Use and the Environment* (5th Ed.) by Hinrichs and Kleinbach. Each unit covers a series of topics as shown in Table 5.1. Some topics are further discussed using documentary DVDs and pre- and postunit questionnaires known as GC dialogues (33). These dialogues help to share student viewpoints and perspectives of green chemistry topics in the classroom.

Service-learning activities that can provide a meaningful community-based learning experience for students have been designed based on the major units discussed in class. These activities provide an opportunity to explore real-life applications based on community needs as students develop intellectually and become more skilled. The community learning projects can facilitate the development of skills that place students at an advantage in their future careers by building their research ability, communication skills, and experience. So far, two community-based service-learning projects have been developed for this course. Considering there is no separate laboratory, only one project is implemented in a given semester. Project 1 correlates with the water unit, and Project 2 correlates with the energy unit of the course. These projects clearly help students understand how aspects of the world around us are impacted by the course themes and are discussed in detail below.

Table 5.1 Green chemistry course units and topics.

Unit	Topics	Supplemental DVDs and Green Chemistry Dialogues (GCD)
Introduction to Green Chemistry	**Principles of Green Chemistry**: Metrics, waste production and prevention, life-cycle assessment	Rachel Carson's Silent Spring (GCD1)
Air	**The Air We Breathe:** Air quality, pollutants and risk assessment **The Chemistry of Global Climate Change**: Ozone, CFCs, greenhouse gases and global warming, Kyoto Protocol	Global Warming: What's up with the Weather? (GCD2)
Water	**Water for Life:** Properties and contamination problems, purification strategies and pollution prevention **Neutralizing the Threat of Acid Rain:** pH of rain water, ocean acidification, photochemical smog, industrial smog	GCD3
Energy	**Energy from Combustion:** Introduction to fossil fuels, gasoline, and biofuels **Energy from Renewable (Noncombustion) Sources:** solar, wind, hydrothermal, and geothermal **The Fires of Nuclear Fission and Electron Transfer:** Radioactivity, half-life, weapons connection, nuclear waste	Solar Energy: Saved by the Sun (GCD4)
Polymers and Plastics	Synthesis, recycling methods, biodegradable plastics	Addicted to Plastics (GCD5)

5.4 Community-Based Learning Project 1: Water quality analysis of the Genesee River

In Community-Based Learning Project 1 (CBLP1), introduced in Spring 2017, students joined with a community partner from the Genesee River Watch to carry out water quality analysis of the river. This organization is dedicated to adopting initiatives in order to reduce pollution of the Genesee River water. The organization works toward achieving safe and healthy use of the river while promoting economic development. One of the main tasks of Genesee River Watch is to frequently analyze the water and to take any action to improve its quality. The organization connects with many local institutions for frequent collection of data to monitor water quality.

Students gathered water from a designated site identified as important by Genesee River Watch for the analysis. Results were shared at the Service-Learning Symposium at SJFC as a poster presentation and with the community partner in a written report. This project was expected to enrich student learning through application of chemical principles and laboratory skills. The six specific learning goals of the CBLP1 are listed in Table 5.2.

5.4.1 Description of the CBLP1

The Genesee River is one of the main nature attractions of the city of Rochester. This river is one of the few in the world that runs north, extending from Potter County in Pennsylvania through Rochester, NY, and ending as it flows into Lake Ontario. The Genesee has a unique geography as it runs for almost 160 miles and drops 2250 ft through a full range of cities and has three different waterfalls *(34)*. In the city of Portageville the river is referred to as the "Grand Canyon of the East" in Letchworth State Park. The river has also become iconic to the city of Rochester and is one of its few natural scenes. Unfortunately, the river is much underappreciated and not used for many recreational activities due to the pollution in some areas.

Ongoing release of phosphorus from different human activities has been found to be the main reason for water pollution in the Genesee River. Agricultural activities, dairy farming, municipal and industrial wastewater treatment plants, and septic systems have all contributed to these existing problems. Phosphate in natural waters can be directly linked to eutrophication *(35)*. This is the excessive concentration of nutrients in a body of water caused by runoff from different sources *(36)*. The final results of eutrophication are increased aquatic plant growth, oxygen depletion, and vast algal blooms. As a result, river water frequently experiences degradation of ecological, economic, and aesthetic value by restriction of use for fishing, drinking water, industry, and recreation.

The timeline for the CBLP1 is shown in Table 5.3. During the first week of the semester, the director of the Center for Service-Learning and Civic Engagement at SJFC visited the classroom and provided an orientation to service-learning including

Table 5.2 Learning goals of the Community-Based Learning Project 1.

1. To understand the environmental indicators of water quality
2. To be able to link the lecture topics in green chemistry with real-world issues
3. To provide services and information that will benefit Genesee River Watch and the community
4. To become skilled in following protocols of collecting water samples from a river and analyzing them for water quality
5. To develop effective oral and written communication skills through writing reports and presenting results to a broader audience
6. To reflect on the service-learning experience of using one's professional skills to benefit the local community

Table 5.3 Timeline for the Community-Based Learning Project 1.

Week	Activity
1	Student orientation to service-learning and water analysis project by the course instructor and the director of the Center for Service-Learning and Civic Engagement.
2	Orientation by a board member of the Genesee River Watch about issues and community need.
3	Exploring laboratory techniques to analyze water, learn laboratory safety.
4	Water collection: student travel to river sites identified by the Genesee River Watch. Water temperature, pH, dissolved oxygen, phosphate, and transparency measurements are taken at the site.
5–7	Analyzing Genesee River water for phosphate levels.
8–9	Analyzing Genesee River water hardness.
10–11	Discussion of the results with the instructor and students and preparation of a preliminary report. Preparation of a poster to be presented at the Service-Learning Symposium attended by students, faculty, and community partners from all service-learning courses.
12–13	Completion and submission of the final report to Genesee River Watch and the poster presentation at the Service-Learning Symposium

the expectations of all parties. Then, the instructor introduced the community learning project more specifically and illustrated the connection of the project with the lecture unit.

During the following week, a member of the Genesee River Watch visited the class for an introduction. The purpose of this introduction was not only to provide specific information about the project in terms of the current pollution conditions of the river and goals for restorations, but also to raise student awareness of environmental policies and the role that Genesee River Watch and other local organizations play in the implementation of a huge restoration project. In the following classroom meetings, students were incrementally introduced to different water analysis techniques and parameters they would use in the field and in the laboratory. Some students in this course do not major in sciences and have never taken a college-level science course. Therefore, it was important to allocate some time to prepare these students for proper laboratory techniques and safety. Students were asked to read the "EPA-Volunteer Stream Monitoring" report which provided background information and appropriate protocols for water sampling and analysis (37). Students were also provided with detailed experimental procedures at the laboratory which were additionally made available at the course website.

During the fourth week of the semester, students conducted sampling at a river site selected by the community partner near downtown Rochester where a Kodak industrial facility is located (Fig. 5.1). The water sampling was demonstrated by a scientist from the Genesee River Watch and supervised by the instructor.

FIGURE 5.1

Sampling site at the Genesee River.

Students worked in pairs and completed on-site water analysis (Table 5.4). Water temperature and pH were measured and a transparency tube (turbidity tube) provided by the Genesee River Watch was used to determine water turbidity and transparency. The transparency tube is made of clear plastic and marked with a centimeter scale and a white-black pattern (Secchi pattern) painted on the bottom *(38)*. River water was poured into the tube until the white-black pattern disappeared and then released from the bottom valve until the pattern became visible. The depth (centimeters) of water was read from the scale attached to the tube. Dissolved

Table 5.4 Water quality tests performed in the Community-Based Learning Project 1.

Location	Test	Method description
River site	pH	pH meter
	Water temperature	Thermometer
	Turbidity and transparency	Secchi disk and transparency tubes
	Dissolved oxygen	Hach kit
	Phosphate	Hach kit
Laboratory	Phosphate	UV-visible spectrometer
	Water hardness	EDTA titration

oxygen and phosphate concentration were measured at the site using a Hach kit, and the phosphate concentration was compared with that measured by a spectrometric method in the laboratory.

The water was sampled, stored in acid-washed polyethylene containers, and then transferred to the laboratory. There, students worked in pairs and conducted phosphate and water hardness analyses. Whenever possible, a science major was paired with a nonscience major to give them the opportunity to share expertise and interests. An alternative method to quantitatively measure phosphate using a UV-visible spectrometer was given to students by the instructor. Hands-on exposure to instrumentation was deemed important to show students the technology available for chemists today. Students prepared all standards, reagents, samples, and blanks and learned the concepts of absorption and calibrating an instrument. Each student pair took independent measurements for each water sample for better accuracy, and the average was calculated. They learned how to use the UV-visible spectrometer, record data in a data table, use Excel to prepare a calibration curve, and how to determine the sample phosphate concentration using the calibration curve. The instructor checked all their calculations for completeness and accuracy. Phosphate concentrations for the water samples generally ranged from 48 µg/L to 83 µg/L, with concentrations greater than about 17−20 µg/L indicating eutrophic conditions (35).

Hardness of the Genesee River water was then determined using a titration method. This provided students with an opportunity to become familiarized with a different analytical technique. Water containing high concentrations of Ca^{2+} and Mg^{2+} ions is called hard water. This can form large amounts of insoluble calcium carbonate and magnesium carbonate when heated, generating scale inside pipes, teakettles, and water heaters. Water hardness is an important measurement as the Genesee River flows into New York's Lake Ontario coastal waters which are a valuable resource for drinking water. The total amount of Ca^{2+} and Mg^{2+} ions was calculated by titrating the river water samples with EDTA using the Eriochrome Black T indicator. Students learned how to accurately perform a titration, determine the end point, and interpret data. They prepared standard solutions and performed three titrations for the standards for better accuracy. They also performed three titrations for each water sample. According to student results, Genesee River water hardness ranged from 300 mg/L to 550 mg/L which is considered as being hard water.

5.4.2 Assessment and grading of the CBLP1

CBLP1 materials and concepts were incorporated into the water unit test of the course to assess some of the course learning goals. Other assessments included the student laboratory notebook, final paper, poster presentation, and a survey as outlined below. During the first few weeks of the community-based service-learning project, students were trained on how to keep a detailed and organized laboratory notebook, including writing detailed methodologies, tabulating data, recording observations, drawing graphs, and doing calculations accurately while paying attention to units. This approach helped students to improve their notebook organization and ensured better quality control of the data. Students submitted their completed notebook at the end of the project and received their grade.

After the completion of all laboratory experiments, students discussed their raw and calculated data with the instructor and classmates. Analysis of scientific data and written communication of results were important expectations of this service-learning project. Once the accuracy of all data and calculations was confirmed, students were required to submit a formal report. As most students did not have any experience in writing a formal scientific report, detailed report guidelines were provided. The typewritten final formal reports included a cover page, a concise abstract, a detailed experimental section with procedures and tabulated data, a discussion of all results, and a references section. They were submitted to the instructor by each pair of students for grading.

The final presentation of the community-based service-learning project was scheduled during the last week of the semester. Students had the opportunity to learn how to collaboratively prepare a poster and present their findings to a broader audience. This was an important learning goal of the project. To avoid repetition, each pair was asked to focus on one designated analysis of the project and to prepare that portion for the presentation. The service-learning symposium was attended by students from all service-learning courses who were taught during the semester, the faculty members who taught the courses, and all community partners.

The community-based service-learning project was set at 30% of the course overall grade. All students were expected to actively participate in all parts of the project including fieldwork, laboratory work, discussions, presentations, and writing reports. The students were graded based on their participation and professionalism, notebook accuracy and completeness, quality of the final presentation, and the quality of the final report. A detailed rubric was used to grade the final presentation and the paper.

5.4.3 Student feedback for the CBLP1

Students were sent a separate survey on the community-based service-learning experience of this course. All candid responses to the free comment portion of the survey from the community-based learning section were positive. Overall, students agreed that the community-based service-learning enhanced the meaningfulness of their learning and that they were able to apply what they learned in the classroom to a real-world issue. Some noteworthy responses are as follows:

> *It showed me the real life applications of this course, and how we need to protect our environment.*

> *We were able to get real data that was going to the community and not just written in a random lab report to be graded and never looked at again.*

> *It was hands on and helped bring to life what we are actually learning.*

Students also commented on the biggest growth areas as a result of community-based service-learning. Their responses varied from changes in the ways they think about community issues and enhanced academic/professional skills development to interest in continuing community service. Some student comments are as follows:

I learned new lab skills and how to report data. Showed how a chemistry degree is applicable in the real world.

Improved knowledge about the water in my own community that I didn't know or care about before.

Group work. I work better in groups, I gained leadership skills.

One major limitation of the CBLP1 was the availability of a community partner and the need for water quality analysis. However, the Genesee River Watch will continue to monitor the river water quality, and we will continue to be able to use this project in the green chemistry course. Also, trying to complete an experiment or fieldwork within a 90-minute period can be a logistical challenge. Advanced planning and preparation are certainly needed for such situations.

5.5 Community-Based Learning Project 2: Developing a waste-to-energy education module for the middle school curriculum

This project is unique and very different from the water analysis project (CBLP1) described before. It correlates with several major topics of the green chemistry course, including renewable energy, waste prevention, and life-cycle assessment (LCA). Students first assist in the development of an energy education unit based on waste-to-energy (W2E) concepts. This unit is intended to be used in middle school science curricula to provide opportunities for students to experience energy-related problem solving and to improve their scientific literacy on energy topics. The unit is housed at the Board of Cooperative Educational Services (BOCES) community partner located in Monroe County, and the project is supported by an Energy-to-Educate grant program from Constellation Energy Company which provides funding for research projects focusing on energy innovation and energy education. The overall project consists of three phases as listed in Table 5.5.

5.5.1 Description of the CBLP2

Students enrolled in the green chemistry course are responsible for working on Phase 1 of the project. They are given biodiesel as the major topic under the W2E category and asked to explore different biodiesel-related projects that have the potential to be used in a middle school science curriculum. The project directly correlates with the energy unit of the course. Students find interesting ideas based on the

Table 5.5 Three phases of the overall waste-to-energy project.

Phase	Task
1	Explore hands-on activities that can be used in the waste-to-energy education module and share with the community partner (BOCES)
2	St. John Fisher College, BOCES, and Pittsford Calkins Middle School use the activities explored in Phase 1 to develop a middle school science module with a science kit that directly correlates with New York State science core curriculum and learning standards
3	Implementation of the module into the science curricula at Rochester area middle schools

topic, conduct field and/or laboratory investigations to determine feasibility, and write a detailed experimental protocol, predict results, and obtain resources to complete the projects. They then share their project ideas, materials lists, experimental procedures, and worksheets with BOCES. The results are also presented in a poster format at the annual SJFC's Research and Creative Work symposium. These projects are intended to help students to learn the process of scientific discovery, improve scientific literacy and communication skills, and teach concepts around renewable energy and zero waste. The specific learning goals of the project are given in Table 5.6.

This project involves students developing hands-on activities to be used in an energy education unit based on W2E concepts. The developed unit with a science kit will be made available for middle schools through the BoSAT program of the Monroe 1 BOCES (Board of Cooperative Educational Services). BOCES was founded by the New York State Legislature. This public organization provides shared educational programs and services to school districts across large geographic regions of the state (39).

Energy education at all age levels can enhance student understanding of the science and technology needed to address energy issues. Understanding the importance of energy efficiency and renewable energy topics, and gaining ability to compare costs and impacts of a wide variety of energy sources will help students to develop important life

Table 5.6 Learning goals of the Community-Based Learning Project 2.

1. To develop research skills though exploring and investigating a project
2. To be able to link the energy topics from the green chemistry lecture with real-world issues
3. To provide services and information that will benefit BOCES to build a middle school science education unit
4. To become skilled in following laboratory protocols and collecting and analyzing data
5. To develop effective written communication skills through writing a project proposal and a report, and oral communication skills through presenting results to a broader audience

skills and perspectives about these increasingly critical areas *(40)*. It is important now more than ever that younger generations have a broad understanding of energy concepts and how their choices impact the local, national and global energy issue.

One of the challenges of infusing energy education into school-level curricula is the lack of immediate access to innovative practical teaching and learning materials. Research has shown that students' motivation toward learning new areas of science depends upon how material is delivered *(41)*. Inquiry-based science education that incorporates active engagement in student learning along with hands-on activities has proven to be more effective than traditional instructional methods *(42)*. Rochester city schools serve a highly diverse, high need, underserved student population. These schools run with minimum resources and have a lower high school graduation rate. It has been reported that early introduction of science through inquiry-based, hands-on activities connected to timely topics can spark an interest in such students for furthering studies in STEM areas *(43)*. The developed W2E unit with hands-on activities and a science kit can address some of these challenges and provide first-class resources for middle school science teachers. Importantly, BOCES did not have any BoSAT science unit that was related to energy concepts and was in need of adding an energy education unit to the existing library.

The timeline for the CBLP2 is given in Table 5.7. This project was integrated for the first time into the green chemistry course taught in Spring 2018. During the first week of classes, students were introduced to the project and the community partner, BOCES. Students learned about the community need, the "big picture," and were tasked with Phase 1 of the overall project which involved developing hands-on active learning biodiesel-related projects (Table 5.5). The project timeline was designed to stimulate a research-type experience that has ultimate use to the community. Students spent 1−2 weeks outside the class time for reading and exploring project ideas. In 2018, the class as a whole agreed to first prepare biodiesel from used cooking oils. This illustrated how waste such as fryer oils can be converted into useful fuels. The topic was also appropriate for middle school students as it involved materials used in everyday life. Under the biodiesel topic, three major projects emerged as follows:

- Project 1: Investigation of the properties of biodiesel: can we use biodiesel during the winter months in Rochester?
- Project 2: Energy, emission, and LCA of biodiesel
- Project 3: Preparation of biosoap from the by-product of biodiesel synthesis

In 2018, the course had six students enrolled and each pair of students was in charge of one of the projects. Each pair completed a separate literature search and gathered information pertaining to their project. As the energy projects correlated closely with the energy unit of the lecture, students gained basic foundation during the lecture on topics such as energy from combustion and sustainable energy sources. Each group wrote a preliminary proposal which consisted of background

Table 5.7 Timeline for the Community-Based Learning Project 2.

Week	Activity
1	Student orientation to community-based learning project, introduction to BOCES, and discussion of the community need.
2–3	Identification of a project of interest under the topic of biodiesel.
4–5	Literature search and information gathering: Once the project is chosen, students conduct a literature search and gather information pertaining to their projects from journal articles, books, local agency flyers, etc.
6	Project planning and submission of a preliminary proposal for approval: students design the project and write a proposal, including background and significance, hypothesis and project goals, experimental plan with required materials and equipment, any safety concerns, and expected results and literature cited.
7–11	Conduct the experiment(s) for the project, collect and analyze data: students work on the project which includes laboratory and/or fieldwork, multiple experiments, data collection, and analyses.
12–13	Students analyze the results of their projects and prepare a poster presentation and a final report.
13	Dissemination of project results: each pair of students presents their poster at the annual Research and Creative Work symposium at St. John Fisher College. This symposium is attended by SJFC's faculty and staff members, graduate and undergraduate students, and community members.
12–13	Completion and submission of the final report.

information and significance, a hypothesis and project goals, and an experimental plan including required materials, equipment, any safety concerns, expected results, and literature cited. Students presented the preliminary proposal during the class and received feedback from the instructor and peers. After revisions, the final proposal was submitted to the instructor. No student enrolled in the course had written a research proposal before. Providing feedback on the preliminary proposal was essential for improving the quality of student scientific writing.

After receiving the project proposals, materials and reagents were ordered based on the student lists. In the meantime, students were given a safety orientation, an introduction to laboratory glassware and equipment, and training on basic practical techniques. All students in the class initially worked together and prepared five types of biodiesel needed for their projects (Table 5.8).

The biodiesel was prepared via a transesterification reaction of oil with methanol using potassium hydroxide as the catalyst *(44)*. Used cooking oils were filtered and heated to remove residual food particles and water prior to transesterification. The by-product of the reaction (glycerol) was removed by centrifugation and saved to be used for the preparation of biosoap (Project 3).

Table 5.8 Biodiesel samples prepared by the class.

Biodiesel from vegetable oil
Biodiesel from corn oil
Biodiesel from peanut oil
Biodiesel from used cooking oil from a local fast-food restaurant
Biodiesel from used cooking oil from the St. John Fisher College's dining hall

5.5.2 Project 1: Investigation of the properties of biodiesel: Can biodiesel be used during the winter months in Rochester?

The main goal of this project was to determine the viscosity and gelation temperature of biodiesel prepared from various pure and used oils and different biodiesel and petroleum diesel blends. Biodiesel derived from different oils can have different viscosity and gelation temperatures. Understanding these properties is important when using biodiesel as a fuel. Viscosity is the measurement of how quickly a liquid flows and is important in how efficiently fuel works in an engine. Biodiesel made up of fatty acid esters can cause oxidation and polymerization which can lead to viscosity increases in storage and gum formation both in storage and use. High viscosity can cause increased fluid friction, reduced operation temperatures, and low energy efficiency. The gelation temperature is the measurement of when the fuel begins to thicken which affects viscosity and therefore engine efficiency.

According to student data, the biodiesel prepared from the used cooking oil from SJFC's dining hall had the lowest relative viscosity. Both 20% and 50% blends of this biodiesel with petroleum diesel provided a viscosity in the same range of 100% petroleum diesel. Biodiesel prepared from fast food oils had the highest viscosity. Further purification of used fast-food oils before transesterification may be necessary to obtain better results. Out of virgin oils used, biodiesel prepared from peanut oil had the highest viscosity. The developed activities of Project 1 involved simple yet interactive hands-on activities that can be done without using expensive equipment or dangerous chemicals. It also covered important basic science concepts such as viscosity and gel temperatures and showcased their connection to real-world applications. These activities have the potential to be used in a middle school science curriculum.

5.5.3 Project 2: Energy, emission and life-cycle assessment of biodiesel

The major goals of this project were to determine energy content by burning various pure biodiesel and blends of biodiesel and petroleum diesel mixtures. Students used a homemade calorimeter for determining the energy content by burning fuels. The calorimeter was prepared by simply using a soda can equipped with a thermometer and a crucible with a candle wick. During the experiment, a specific amount of water

(50 g) was added to the soda can and the initial temperature was measured. Then, a specific amount of biodiesel (2 g) was added to the crucible, and the candle wick was lit. Once all the oil was burned, the temperature increase was noted. The amount of heat transferred to the water was determined by using the equation $q = mC\Delta t$, where q is the heat in Joules, m is the mass of water in grams, C is the heat capacity of water (4.18 J/g°C), and Δt is the temperature change of the water in degrees Celsius. The heat of combustion was then calculated by dividing q by the mass of the fuel. According to the student results, the biodiesel prepared from the used cooking oil from SJFC's dining hall had the highest heat energy. The amount of energy produced from 25% blend of biodiesel prepared from dining hall oil and petroleum diesel was closer to the energy produced from 100% petroleum diesel. The students also determined the amount of soot left in the crucible by weighing the crucible before and after the experiment. According to their data, petroleum diesel and the biodiesel produced from the fast-food restaurant produced the most soot. Additionally, students developed a LCA for biodiesel and compared it with the LCA of petroleum-based diesel. LCA is an important green chemistry tool used to evaluate the environmental impacts of a product or a process. Using the LCA, interested parties can gain knowledge about where the most important problems lie within a life cycle and find the solutions to address those problems.

The developed activities of Project 2 involved in-house construction of a calorimeter using simple materials and measurement of two important fuel parameters. When the amount of water used and the fuel burnt is consistent for all experiments, the temperature change of water is proportional to the heat of combustion. This simplifies the activity for middle school students by removing the need for using a mathematical formula, making it an exciting experience to learn how everyday materials can be used to construct an important scientific device and to see how fuel is burnt to produce energy.

5.5.4 Project 3: Preparation of biosoap from the by-product of biodiesel synthesis

The major goal of this project was to develop an efficient method to convert waste glycerol into liquid biosoap. This can reduce the amount of waste produced when preparing biodiesel. Students were also meant to compare the efficiency, effectiveness, and cost of biosoap to conventional soap and to investigate the soap usage in SJFC bathrooms.

When preparing biodiesel, a substantial amount of glycerol is produced as a by-product. Glycerol is a major component used in soap preparation. Instead of disposing glycerol as waste, it can be used to prepare soap, increasing the efficiency and economic benefits of the biodiesel preparation process. Students used glycerol produced in biodiesel synthesis from different oils and converted it to biosoap via a saponification reaction. They also prepared soap from commercially available glycerin for comparison. The pH of the soap was balanced and essential oils added for scent. Students also explored the campus-wide soap usage by investigating the

Table 5.9 Poster titles of Community-Based Learning Project 2.

Project	Poster title
1	Can we use biodiesel in Rochester winters? Investigation of the properties of biodiesel
2	How fast can biodiesel burn? Energy content and life-cycle assessment of biodiesel
3	Preparation of biosoap from cooking oil: Used cooking oils from the dining hall to biosoap for campus bathrooms

number of dispensers used in campus bathrooms, how much soap each dispenser held, how much each cost to refill, how many refills were done in a month, and the approximate monthly cost for soap.

With the completion of the laboratory work, the Spring 2018 students prepared three professional quality posters discussing each project and presented them at the annual Research and Creative Work symposium held at SJFC during April (Table 5.9).

Students also displayed their biodiesel and biosoap samples at the symposium (Fig. 5.2). The symposium was attended by hundreds of undergraduate students, faculty and staff, community members, and high school students from two local high schools. This assignment helped to enhance students' scientific literacy and communication skills by presenting their results to a broader audience and promoted civic pride and responsibility by educating a large group of people about green chemistry and energy-related topics. The project results were also submitted as a final report for the instructor to share with BOCES.

Even though Phase II (converting the W2E activities into a BoSAT science module) and Phase III (implementing the science module in middle school curriculum) were not directly part of the CBLP2 of this course, it is worth noting how the project was completed after Phase I. Through the CBLP2 itself, students contributed to the community partner by developing materials for the middle school science unit. Students investigated the feasibility of different experiments and provided detailed procedures for biodiesel synthesis, density and gelation experiments, and calorimetric experiments. They also provided background information for those experiments that can be included in a teacher's manual. Additionally, students prepared data sheets and supply lists that can be used with the science module and the kit.

During Phase II (once the course was completed in Spring 2018), SJFC, Pittsford Middle School, and BOCES worked through the summer converting the projects into a middle school science module. The developed science curriculum includes 10 lessons based upon W2E concepts and correlates the curriculum directly with New York state science core curriculum and next-generation learning standards. Extra lessons were added to introduce major concepts such as density and calorimetry before doing the experiments. The unit also includes a science kit with materials and

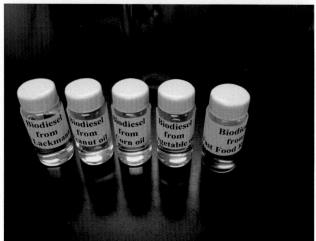

FIGURE 5.2

Biodiesel and biosoap prepared in the Community-Based Learning Project 2.

equipment for hands-on activities for an entire class of students, a student activity book with data sheets and reading materials for students, a teacher manual including background information on renewable energy topics, detailed plans for 10 lessons, procedures for inquiry-based activities, and pre- and post-assessments. The developed module with the science kit is made available to all middle schools in the Rochester area through BOCES. Since BOCES refurbish and maintain their kits, this education unit will serve hundreds of middle school students for years to come.

For Phase III, the curriculum was first implemented at Pittsford Central School District (PCSD) and integrated into the sixth grade science curriculum at Pittsford

Calkins Road Middle School during Fall 2018. Additionally, two students (Student 1: BS Inclusive Adolescence Education/English; Student 2: BS Biology/Sustainability minor) from the green chemistry course were recruited to participate as teacher assistants in W2E unit implementation at Pittsford Calkins Road Middle School. Both these students are planning on entering a teaching career after graduation. This was a rewarding opportunity for both students, and it gave them joy and pride to see what they helped develop being used to educate the younger generation in energy and green chemistry topics.

5.5.5 Assessment and grading of CBLP2

The students were assessed based on pre- and post-work quizzes, quality, completeness, and the accuracy of the project proposal, notebook, final report, and poster. The grade portion for the community-based learning project was 35% of the course overall grade. All students were expected to actively participate in all parts of the project including laboratory work, discussions, presentations, and writing reports. Students were given detailed guidelines on preparing a project proposal, final paper, and poster presentation. In all cases, their first drafts were reviewed, and feedback was provided whenever needed to improve the quality of the assignment. Additionally, their laboratory notebooks were frequently checked for accuracy and completeness. A detailed rubric was used to grade the final presentation and the paper. Student knowledge on energy topics were tested with a quiz before starting the W2E projects. The same quiz was administered after completing the projects. Their percentage improvement from prequiz to postquiz was calculated, and it was apparent that each student had improved (by between 5% and 45%).

5.5.6 Student feedback on CBLP2

Students highly rated the community-based learning experience of this course in their general evaluation forms and the separate evaluation collected for the community learning project. Written responses highlighted a meaningful learning experience, real-world application of course materials, education of the younger generation, academic growth, and enhancing teamwork and scientific communication skills. Some noteworthy responses are as follows:

Knowing that the project we developed will be passed on to younger students in middle school was a very unique and fun aspect of this course

Learning how to prepare a poster and being able to present our work at the symposium was great

I've always had an idea of a career I wanted to pursue after graduation. However, this course and the energy project helped me finalize my career choice

The two undergraduate students who participated in both developing the CBLP2 and teaching it at the Calkins Road Middle School were given surveys to describe their experience in the whole process. One student noted that: "my background with teaching centers around English Language Arts, so teaching a science course was new and exciting to tackle. I have learned that teaching science courses can be challenging because not only is the curriculum incredibly dense, but the kids need to be able to stay engaged, keep up a fast learning pace, and be able to comprehend complicated material and vocabulary. Having hands-on activities in timely topics definitely helps. It was rewarding to see the unit I helped developing put into action and the excitement of these kids' eyes. For this experience, the students made it easy and enjoyable. I have learned just as much from them as they have learned from me through this experience."

Importantly, one limitation of this type of community-based learning project is the need to establish a relationship with a local school district. Sometimes, already existing district curricula may not allow or have room to introduce new modules. This limitation may be overcome by working with an after-school science club.

5.6 Conclusion

Two community-based learning projects have been designed and integrated into a green chemistry course. The projects were designed to stimulate research-type experiences that have ultimate use to the community. In the first project, working collaboratively with a community partner, students analyzed the water quality of the Genesee River. Results were shared at a symposium as a poster presentation and with the community partner in a written report. Green chemistry students developed curricular activities based on W2E concepts for a middle school science unit in the second project. They investigated the feasibility of different experiments and provided detailed procedures, background information, and supply lists for biodiesel synthesis, density and gelation experiments, and calorimetric experiments. The developed activities were later converted into a science unit with a supply kit and implemented in a middle school science curriculum. These projects enriched student learning through application of chemistry principles and laboratory skills based on a community need.

Acknowledgments

I wish to thank all the students enrolled in the green chemistry course and the following people: Dr. Lynn Donahue (Director, Center for Service-Learning and Civic Engagement at SJFC); Dr. Michael Boller (Coordinator, Center for Sustainability at SJFC); Mike Hugh (Director, Genesee River Watch); Mary Humphreys (Middle School ELA/Science Teacher, Pittsford School District); Steve Orcutt (Science Program Director, BOCES and Constellation); and the Energy-to-Educate grant program from Constellation for funding.

References

1. Towns, M. H. Kolb for Chemists: David A. Kolb and Experiential Learning Theory. *J. Chem. Educ.* **2001,** *78,* 1107–1109.
2. Jacoby, B. *Service-Learning in Higher Education: Concepts and Practices;* Jossey-Bass: San Francisco, CA, 1996.
3. Owens, T. R.; Wang, C. Community-based Learning: A Foundation for Meaningful Educational Reform. In *School Improvement Research Series.* Education Northwest: Portland, OR, 1996.
4. Cartwright, A. Science Service Learning. *J. Chem. Educ.* **2010,** *87,* 1009–1010.
5. Cavinato, G. Service Learning in Analytical Chemistry: Extending the Laboratory Into the Community. In *Active Learning: Models from the Analytical Sciences,* Vol. 970, Mabrouk, P. A., Ed.; American Chemical Society: Washington, DC, 2007; pp 109–122.
6. Roberts-Kirchhoff, E. S.; Benvenuto, M. A.; Mio, M. J. Service Learning, Chemistry, and the Environmental Connections. In *Service Learning and Environmental Chemistry: Relevant Connections,* Vol. 1177, Roberts-Kirchhoff, E. S., Benvenuto, M. A., Mio, M. J., Eds.; American Chemical Society: Washington, DC, 2014; pp 1–4.
7. Nickel, A.-M. L.; Farrell, J. K. Using Service Learning to Teach Students the Importance of Societal Implications of Nanotechnology. In *Service Learning and Environmental Chemistry: Relevant Connections,* Vol. 1177, Roberts-Kirchhoff, E. S., Benvenuto, M. A., Mio, M. J., Eds.; American Chemical Society: Washington, DC, 2014; pp 123–134.
8. Najmr, S.; Chae, J.; Greenberg, M. L.; Bowman, C.; Harkavy, I.; Maeyer, J. R. A Service-learning Chemistry Course as a Model to Improve Undergraduate Scientific Communication Skills. *J. Chem. Educ.* **2018,** *95,* 528–534.
9. Esson, J. M.; Stevens-Truss, R.; Thomas, A. Service-learning in Introductory Chemistry: Supplementing Chemistry Curriculum in Elementary Schools. *J. Chem. Educ.* **2005,** *82,* 1168–1173.
10. Hatcher-Skeers, M.; Aragon, E. Combining Active Learning With Service Learning: A Student-driven Demonstration Project. *J. Chem. Educ.* **2002,** *79,* 462–464.
11. Heider, E. C.; Valenti, D.; Long, R. L.; Garbou, A.; Rex, M.; Harper, J. K. Quantifying Sucralose in a Water-treatment Wetlands: Service-learning in the Analytical Chemistry Laboratory. *J. Chem. Educ.* **2018,** *95,* 535–542.
12. Kammler, D. C.; Truong, T. M.; VanNess, G.; McGowin, A. E. A Service-learning Project in Chemistry: Environmental Monitoring of a Nature Preserve. *J. Chem. Educ.* **2012,** *89,* 1384–1389.
13. Kesner, L.; Eyring, E. M. Service-learning General Chemistry: Lead Paint Analyses. *J. Chem. Educ.* **1999,** *76,* 920–923.
14. Draper, A. J. Integrating Project-based Service-learning Into an Advanced Environmental Chemistry Course. *J. Chem. Educ.* **2004,** *81,* 221–224.
15. Glover, S. R.; Sewry, J. D.; Bromley, C. L.; Davies-Coleman, M. T.; Hlengwa, A. The Implementation of a Service-learning Component in an Organic Chemistry Laboratory Course. *J. Chem. Educ.* **2013,** *90,* 578–583.
16. Harrison, M. A.; Dunbar, D.; Lopatto, D. Using Pamphlets to Teach Biochemistry: A Service-learning Project. *J. Chem. Educ.* **2013,** *90,* 210–214.
17. Goeden, T. J.; Kurtz, M. J.; Quitadamo, I. J.; Thomas, C. Community-based Inquiry in Allied Health Biochemistry Promotes Equity by Improving Critical Thinking for Women

and Showing Promise for Increasing Content Gains for Ethnic Minority Students. *J. Chem. Educ.* **2015,** *92,* 788–796.

18. Kirchhoff, M. M. Topics in Green Chemistry. *J. Chem. Educ.* **2001,** *78,* 1577.

19. Cannon, A. S.; Levy, I. J. The Green Chemistry Commitment: Transforming Chemistry Education in Higher Education. In *The Promise of Chemical Education: Addressing Our Students' Needs,* Vol. 1193, Daus, K., Rigsby, R., Eds.; American Chemical Society: Washington, DC, 2015; pp 115–125.

20. Uffelman, E. S. News From Online: Green Chemistry. *J. Chem. Educ.* **2004,** *81,* 172–176.

21. Hjeresen, D. L.; Schutt, D. L.; Boese, J. M. Green Chemistry and Education. *J. Chem. Educ.* **2000,** *77,* 1543–1547.

22. Braun, B.; Charney, R.; Clarens, A.; Farrugia, J.; Kitchens, C.; Lisowski, C.; Naistat, D.; O'Neil, A. Completing our Education. Green Chemistry in the Curriculum. *J. Chem. Educ.* **2006,** *83,* 1126–1129.

23. Barcena, H.; Tuachi, A.; Zhang, Y. Teaching Green Chemistry With Epoxidized Soybean oil. *J. Chem. Educ.* **2017,** *94,* 1314–1318.

24. Young, J. F.; Peoples, R. ConfChem Conference on Educating the Next Generation: Green and Sustainable Chemistry-Education Resources From the ACS Green Chemistry Institute. *J. Chem. Educ.* **2013,** *90,* 513–514.

25. Prescott, S. Green Goggles: Designing and Teaching a General Chemistry Course to Nonmajors Using a Green Chemistry Approach. *J. Chem. Educ.* **2013,** *90,* 423–428.

26. Timmer, B. J. J.; Schaufelberger, F.; Hammarberg, D.; Franzen, J.; Ramstrom, O.; Diner, P. Simple and Effective Integration of Green Chemistry and Sustainability Education Into an Existing Organic Chemistry Course. *J. Chem. Educ.* **2018,** *95,* 1301–1306.

27. Dicks, A. P.; Batey, R. A. ConfChem Conference on Educating the Next Generation: Green and Sustainable Chemistry-Greening the Organic Curriculum: Development of an Undergraduate Catalytic Chemistry Course. *J. Chem. Educ.* **2013,** *90,* 519–520.

28. Manchanayakage, R. Designing and Incorporating Green Chemistry Courses at a Liberal Arts College to Increase Students' Awareness and Interdisciplinary Collaborative Work. *J. Chem. Educ.* **2013,** *90,* 1167–1171.

29. Gross, E. Green Chemistry and Sustainability: An Undergraduate Course for Science and Nonscience Majors. *J. Chem. Educ.* **2013,** *90,* 429–431.

30. Cummings, S. D. ConfChem Conference on Educating the Next Generation: Green and Sustainable Chemistry-Solar Energy: A Chemistry Course on Sustainability for General Science Education and Quantitative Reasoning. *J. Chem. Educ.* **2013,** *90,* 523–524.

31. Kennedy, S. A. Design of a Dynamic Undergraduate Green Chemistry Course. *J. Chem. Educ.* **2016,** *93,* 645–649.

32. *Mission and vision of St. John Fisher College.* https://www.sjfc.edu/about/mission-strategic-plan/.

33. (a) Ian Connacher, Author*Addicted to Plastic (DVD);* Bullfrog Films, 2008.
    (b) Steven Latham, Director*Solar Energy (DVD);* WGBH Boston Video, 2007.
    (c) Jon Palfreman, Director*Global Warming (DVD);* WGBH Boston Video, 2000.
    (d) Neil Goodwin, Producer*Silent Spring (DVD);* WGBH Boston Video, 2007.

34. *The Genesee River Watch.* https://www.geneseeriverwatch.org/.

35. Robertson, D. M.; Rose, W. L.; Saad, D. A. *Water Quality and the Effects of Changes in Phosphorus Loading to Muskellunge Lake, Vilas County, Wisconsin.* U.S. Geological Survey Water-Resources Investigations Report 03-4011, 2003.

36. Kharat, S.; Pagar, S. D. Determination of Phosphate in Water Samples of Nashik District (Maharashtra State, India) Rivers by uv-visible Spectroscopy. *E-J. Chem.* **2009,** *6* (S1), S515–S521.

37. *Monitoring and Assessing Water Quality, Volunteer Stream Monitoring: A Methods Manual.* http://www.epa.gov/owow/monitoring/volunteer/stream/.

38. Dahlgren, R.; Nieuwenhuyse, E. V.; Litton, G. Transparency Tube Provides Reliable Water-quality Measurements. *Calif. Agric.* **2004,** *58,* 149–153.

39. *The BOCES.* http://www.monroe.edu.

40. Pipere, A.; Grabovska, R.; Jonane, L. Inspiring Teachers for Energy Education: An Illustrative Case Study in the Latvian Context. *J. Teacher Educ. Sust.* **2010,** *12,* 37–50.

41. National Research Council. *Taking Science to School: Learning and Teaching Science in Grades K-8 (STEM Education);* National Academy Press: Washington, DC, 2007.

42. Harris, C. J.; Rooks, D. L. Managing Inquiry-based Science: Challenges in Enacting Complex Science Instruction in Elementary and Middle School Classrooms. *J. Sci. Teacher Educ.* **2010,** *21,* 227–240.

43. Trna, J.; Trnova, E.; Svobodova, J. *Inquiry-based Science Education and Experiments.* http://www.cen.uni.wroc.pl/Pliki/Wydawnicza/19_wybrane__teksty/10_trna.pdf.

44. *Loyola Biodiesel Program.* https://www.luc.edu/sustainability/initiatives/biodiesel/.

Promoting political and civic engagement in a nonmajor sustainable chemistry course

Loyd D. Bastin, PhD [1], Andrea E. Martin, PhD [2]

Professor, Departments of Chemistry and Biochemistry, Widener University, One University Place, Chester, PA, United States[1]; Associate Professor, Department of Chemistry, Widener University, One University Place, Chester, PA, United States[2]

6.1 Introduction

In *Greater Expectations: A New Vision for Learning as a Nation Goes to College (1)*, the American Association of Colleges and Universities indicated that general education should promote student learning through the development of critical thinking and civic engagement. A key component of general education is science education. Every college student should understand the scientific process. Traditionally, chemistry education has focused only on discipline-specific, technical knowledge. More recently, chemistry educators have developed courses and contents that strive to meet specific goals using real-world issues and civic engagement that engage nonmajors in chemistry content within their community *(2,3)*.

Currently, there are a number of examples of college courses focused on environmental issues that affect local and/or global communities. For example, Metzger developed an environmental justice course that brings together liberal arts students and incarcerated men *(4)*. The course allows students to explore how human activities have environmental consequences and how cultural, political, and economic factors affect environmental policy-making. Another faculty member has developed a number of chemistry courses that explore arsenic in food and water, promoting awareness and science literacy through formal and informal learning *(5)*.

Boyer argues that the mission of higher education should be to educate students for a life of responsible citizenship rather than solely educating students for a career *(6)*. His argument rings particularly true at Widener University where this is the heart of our institutional mission. As a comprehensive university, Widener merges professional education with the liberal arts. Widener and many other colleges and universities embed civic engagement into general education and discipline-specific courses. The goal of our civic engagement work is to improve the quality of living in our communities through political and nonpolitical processes *(7)*. Civic engagement takes many forms including voting, volunteering, political action, and staying informed *(8)*. Most institutions of higher education include an obligation to prepare undergraduates to be civically responsible in their mission and/or goals *(9)*. These goals are typically accomplished through extracurricular activities (usually volunteering) and curricular activities (often through service-learning courses). While volunteering is an important activity that contributes significantly to personal development, academic service-learning activities have significant benefits to the student, teacher, and community *(9—11)*.

Service-learning has been identified by Kuh as a high-impact educational practice *(12)*. It is a structured teaching and learning experience that meets identified needs in the community with explicit learning objectives, preparation, and reflection *(13)*. The value of service-learning to the teacher, community, and the learner has been well documented *(14,15)*. The activities have a positive effect on the learner's beliefs and values toward service and community *(14—19)* and on critical thinking and the ability to integrate theory *(16,20—25)*. As a result of these positive outcomes, examples of service-learning in chemistry have been published in the pedagogical literature. These examples embed into the course objectives as follows: (1) demonstration or generation of materials by students in schools or the community *(26—33)*; (2) teaching primary and secondary school students or helping secondary school teachers with curriculum revision *(34—36)*; or (3) application of chemical analysis techniques to environmental issues in the community *(15,37—45)*.

Another pedagogical tool that successfully integrates civic engagement into the chemistry curriculum is the SENCER (Science Education for New Civic Engagement and Responsibilities) approach. SENCER's core goals are to (1) encourage more student interest and engagement in STEM; (2) help students connect STEM learning to their other studies; and (3) strengthen students' understanding of science and their capacity for responsible work and citizenship *(46)*. SENCER aims to enhance student learning by connecting science content with real-world issues through civic engagement *(47)*. Evidence from Student Assessment of their Learning Gains (SALG) data shows that students in SENCER classes are more engaged in STEM than students in non-SENCER courses, they are more likely to integrate knowledge, and they have greater learning gains in critical thinking skills *(48)*. The SENCER community has published a variety of chemistry examples in three American Chemical Society (ACS) books *(49—51)*.

A number of these examples use environmental issues to civically engage students in the course, with several designed for science majors. Latch described a collaborative, interdisciplinary service-learning project between ecology, organic chemistry laboratory, and instrumental analysis courses that involves field sampling, chemical analyses, and research-based experiments *(52)*. Viviano et al. designed a service-learning course that connects students with high school environmental educators and a local nonprofit environmental agency to address environmental issues in their local community *(53)*. Eichler embedded a long-term environmental project into a general chemistry course for science and nonscience majors where the students measure ground-level ozone concentrations as part of a long-term air quality study *(54)*.

Additional examples were designed for nonscience majors as a way to engage these students with scientific content and showcase the relevance of science in the general education of all students. Railing designed an "Introduction to Environmental Issues" course around six global ecological challenges using current real-world case studies. The course consisted of a service-learning project to build a vertical hydroponic growing wall for a local food pantry *(55)*. Odenbrett described the design of an inquiry-based approach to connect chemistry topics to real-life issues by following the pathways of phosphorus runoff from northwestern Ohio farm fields to Lake Erie. A civic engagement piece is added to data collection using the GLISTEN (Great Lakes Innovative Stewardship Through Education Network) project that creates clusters of STEM faculty to collaboratively address Great Lakes ecosystem challenges *(56)*. These clusters connect students and faculty to local public and private nonprofit environmental agencies. Maguire and da Rosa described two environmentally focused general education core courses that include civic engagement activities *(57)*. Their "Introduction to Environmental Chemistry" course uses *Chemistry in Context* to relate relevant chemistry knowledge to sustainability topics. The civic engagement activities included promoting a better campus paper-recycling program and campus garden workdays. The second course focuses on the science, history, and sociology of climate change using case studies. Bachofer collaboratively designed a general education science course with sociology faculty that studies the redevelopment of a superfund site. A faculty learning community approach was used to keep courses connected and on task for the civic engagement activity. The civic engagement component involved a trip to learn about city development, field sampling, and X-ray fluorescence soil screening experiments *(58)*. Davis and Fisher described a science and global sustainability course for nonscience majors using sustainability issues as a set of interconnected real-world problems to teach science content *(59)*.

Several non-SENCER examples embed civic engagement into green chemistry and sustainability courses. Gurney and Stafford described an upper-level capstone seminar for science and engineering students that features the Presidential Green Chemistry Challenge award projects as course materials *(60)*. The students organized an event for the campus community to highlight green chemistry research. They have also described two nonscience courses that connect green chemistry and civic engagement. At Simmons College, a general education honors course was designed that mapped the *Chemistry in Context (61)* topics with real-world cases from *Watershed (62,63)* readings. The students researched green consumer products and submitted weekly articles to the student newspaper to inform the campus community. They also created a green "What to Bring to College" guide for incoming students. Finally, the students initiated and championed the creation of a university-wide sustainability committee. These experiences inspired Gurney and Stafford to design a learning-community course that explores the intersection of green chemistry and environmental ethics. The last two weeks of the course are dedicated to student-designed civic engagement activities *(60)*. Bouvier-Brown described three exercises for chemistry courses that integrate atmospheric chemistry and environmental justice *(64)*. This approach brings the focus of education onto the humans affected by environmental problems. The curriculum uses real-world and student-collected air pollution datasets to analyze a variety of air pollution conditions. While an explicit civic engagement activity is not provided, an instructor could use these assignments to guide the creation of similar datasets for their local area and integrate civic engagement activities through local environmental groups.

6.2 Widener University

Widener University is a leading metropolitan university in the Philadelphia area. The university consists of three campuses: the main campus in Chester, PA, and law schools in Wilmington, DE, and Harrisburg, PA. The undergraduate student body numbers about 3300 students *(65)* with about 90% of them being full-time on the Chester campus. The largest undergraduate enrollments are found in the College of Arts & Sciences and the School of Nursing, followed by the Schools of Business Administration and Engineering. The university also has sizable graduate enrollments in the Schools of Human Service Professions, Business, Nursing, and Law.

6.3 Civic engagement at Widener

The university's mission statement highlights the importance of civic engagement *(66)*.

Here at Widener, a leading metropolitan university, we achieve our mission by creating a learning environment where curricula are connected through civic engagement. Our mission at Widener includes the following tenets:

- *We lead by providing a unique combination of liberal arts and professional education in a challenging, scholarly, and culturally diverse academic community*
- *We engage our students through dynamic teaching, active scholarship, personal attention, and experiential learning*
- *We inspire our students to be citizens of character who demonstrate professional and civic leadership*
- *We contribute to the vitality and well-being of the communities we serve*

Former President James T. Harris III recognized the needs of Chester and created an opportunity for the university to create a bridge (both figurative and literal) with the community *(67)*. One of the first steps was to create a service-learning fellowship program for faculty members. Faculty were awarded release time to create courses that would offer students a learning experience in the Chester community. Service-learning classes are intended to provide a reciprocal benefit to both the student and the community. Community partners may include public schools, nonprofit organizations, and agencies. A key requirement of service-learning courses, which receive a special transcript designation, is the opportunity for students to reflect on their experience (pre-service, during service, and post-service) through class discussion and/or written reflections.

Service-learning is one of several programs administered by the university's Office of Civic and Global Engagement. Widener offers undergraduate scholarships through the Presidential Service Corps/Bonner Leader program. Students in this program are required to volunteer 300 h annually in community service in the city of Chester. In partnership with Habitat for Humanity, Widener offers Alternative Spring Break opportunities in which teams of students and faculty work on construction projects in areas of need throughout the country. Widener's nursing, physical therapy, and law students offer low-cost or pro bono clinics in their communities. The university maintains the Widener Child Development Center which provides nursery, preschool, and kindergarten experiences for children of Chester residents and Widener faculty and staff. In 2006, Widener launched the Widener Partnership Charter School. Originally offering kindergarten and first grade, the school has grown to include grades K-8.

Through the Political Science Department in the College of Arts and Sciences, Widener also engages students in political action. Widener works with the nonprofit Project Pericles *(68)* to embed social responsibility and civic engagement in the curriculum. In 2012, university students and faculty formed a political action committee (PAC), called College Students Concerned by College Costs. That group has participated regularly in the annual Association of Independent Colleges and Universities of Pennsylvania (AICUP) Student Aid Advocacy Day ("Lobby Day") at the State Capitol in Harrisburg. There, students have the opportunity to meet with state legislators and staff to discuss issues of concern.

6.4 Sustainability at Widener

Sustainability forms the nucleus of one of the goals articulated in the university's strategic plan, "Vision 2021." In 2015, Professor James May (Widener University, Delaware Law School) was named the Chief Sustainability Officer for the University. His first annual report *(69)* detailed work done to infuse sustainability into curricula, research, operations, and student life. He envisioned a "three pillars" model consisting of campus, classroom, and community engagement. Professor May created a campus Sustainability Council consisting of faculty, staff, students and community members in order to oversee efforts in these areas. On the main campus, the newer buildings have LEED certification, we are switching to LED lighting, and we use geothermal energy to heat and cool seven buildings. Students have led efforts to reduce food waste and promote the use of sustainably-farmed products in the cafeteria. Our on-campus coffee shops serve "WUBrew," an organic shade-grown coffee produced in Las Lajas, Costa Rica. The university maintains a sustainability hub in Costa Rica where interdisciplinary groups of students and faculty have worked with the local coffee farmers to make this a successful enterprise. Professor May also instituted an annual Sustainability Expo, which has included internal and external speakers, panel discussions, student posters, and displays. This is typically held in the spring, around the time of Earth Day.

We have recently added minors in Sustainability Science and Sustainability Management to a curriculum that features sustainability across many programs, from science and engineering to business, law, and the humanities. In August 2015, the university offered its first workshop about infusing sustainability into the curriculum. Sponsored by former Dean Sharon Meagher and modeled after a similar program at the University of Scranton, the workshop included faculty from nursing, hospitality management, psychology, English, chemistry, and business. Presentations were given by other science, humanities, and law faculty and a member of the Pennsylvania Environmental Council; the group also received an environmental justice tour of Chester, led by the Reverend Dr. Horace Strand (discussed in Section 6.7). By the end of the workshop, participants had developed course syllabi and committed to reporting back on the outcomes of their classes.

The infusion of sustainability into the chemistry curriculum started in 2007, when Loyd Bastin began "greening" the organic chemistry laboratories. Rather than a top-down approach, he instead involved both research students and organic chemistry students in analyzing experiments based on the 12 Principles of Green Chemistry *(70)*. Students in the organic laboratory were asked to suggest improvements to the experiments, which were tested by undergraduate research students, and their results were incorporated into the next year's class *(71)*. Based on this work, Bastin championed the expansion of green chemistry throughout the curriculum. In addition, several faculty members began to incorporate green chemistry into their research programs. In 2014, the Widener Chemistry Department became the first school in Pennsylvania to sign the Green Chemistry Commitment *(72)*, a program organized by the nonprofit organization Beyond Benign *(73)*. The members of the Green Chemistry Commitment share best practices and develop resources through working groups and conference symposia. Green chemistry is now embedded throughout the department's curriculum and is part of the department's mission statement and student learning goals. The Chemistry Department also continuously looks to improve the sustainability of our daily operations. Recently, the department replaced all faucet aspirators with recirculating water aspirator pumps, and 50 exhaust hoods were replaced with recirculating hoods. These improvements to the operations infrastructure significantly reduce our water and energy consumption and greatly decrease the pollution of the environment with carbon dioxide and volatile chemicals.

6.5 Chester

Central to Widener's civic engagement mission is our location. The City of Chester lies on the Delaware River, a few miles southwest of Philadelphia. Settled as "Upland" by the Swedes in 1644, it is the oldest city in Pennsylvania *(74)*. In 1682, William Penn landed in the city and changed its name to Chester. The city grew to be a prosperous manufacturing hub based on shipbuilding, metal manufacturing, and textiles industries. During World War I, these industries thrived and the population grew from 38,000 to 58,000 between 1910 and 1920. While ship-building declined after World War I, other manufacturing, such as a Ford automobile assembly plant, stepped in. A second boom occurred during World War II, and the city's population reached 66,000 in 1950 *(74)*. Soon after, however, industries began to abandon the city, resulting in loss of jobs and subsequent migration of people to surrounding areas. In 2010, the population stood at fewer than 34,000 people, less than 20% of whom were white *(75)*. The 2013−17 unemployment rate was 18.1%, and the median household income was about $30,000; over 40% of the households were reported as using food stamp/SNAP benefits *(76)*. These statistics are staggering compared to the same data for the state of Pennsylvania where the unemployment rate was 6.5%, the median household income was about $57,000, and 13% of the households were reported to use food stamp/SNAP benefits *(77)*. Today, less than 10% of the workforce is employed in manufacturing, with the bulk of employment in healthcare, retail, food service, and education *(78)*.

It has been recognized for decades that Chester is a case study in environmental injustice *(79,80)*. The city is home to the notorious Wade Superfund Site, originally a rubber recycling facility, where thousands of gallons of toxic waste had been stored illegally prior to a disastrous fire in 1978 *(81)*. After remediation, the site was deleted from the National Priorities List in 1989, and it is now a parking lot for a profes-sional soccer league stadium in Chester. The Wade site is just one of many environ-mental issues in the city. A residential area along the Delaware River is the site of one of the largest trash-to-steam generators in the country, accepting waste from a wide portion of the Middle Atlantic region. Covanta identifies this facility as their largest energy-from-waste operation worldwide, processing over 1.3 million tons of municipal solid waste and producing over 645,000 MWh of electricity annually *(82)*. Next door is the county's sewage treatment plant, DELCORA, which processes over 100 million gallons of wastewater daily from Delaware County *(83)*. The ef-forts of the community to organize in opposition to serving as a dumping ground for wealthy neighboring suburbs are detailed in the award-winning 1997 documen-tary, *Laid to Waste (84)*. A community activist group, Chester Residents Concerned for Quality Living (CRCQL), filed a civil rights suit in 1993 to stop the opening of a large infectious waste facility in an adjacent lot to the Covanta and DELCORA fa-cilities. This facility had received a permit from the Pennsylvania Department of Environmental Resources despite exceeding regulatory limits by a factor of 10 *(79)*. Two years later, the Pennsylvania Supreme Court revoked the permit based on Title VI of the Civil Rights Act of 1964, confirming that racism had played a role in the location of the waste treatment facilities.

The 2013 air quality index (AQI) for Chester was 119, significantly higher than the United States as a whole, which was 75. Specific substances that exceeded national averages included nitrogen dioxide and lead *(85)*. Chester is a significant contributor to the fact that Delaware County was listed as #20 on the "Top 101 Counties with the Highest Particulate Matter ($PM_{2.5}$) Annual Air Pollution Readings in 2012 ($\mu g/m^3$)" *(85)*. Air pollution is a likely factor in Chester's high asthma rate (18% adult asthma rate based on 2010 statistics, compared to a state and national rate of 13.3%) *(86)*. The air quality issues and related health effects are undoubtedly related to Chester's close proximity to interstate I-95 and the heavy truck traffic generated by Covanta's waste-to-energy facility.

The Chester Environmental Partnership (CEP) is an offspring of the CRCQL and was created to bring together Chester city residents and local businesses. Formed in 2005 under the leadership of Reverend Dr. Horace Strand *(87)*, it is a grassroots nonprofit organization comprising concerned members of the community; representatives of local businesses including Covanta, DELCORA, and Kimberly—Clark; members of the Pennsylvania Environmental Resources Council; staff from the Pennsylvania Department of Environmental Protection; and representatives from Widener University and the University of Pennsylvania. The CEP aims to promote environmental sustainability and social justice in the Chester area. The group has worked effectively with local industry to reroute truck traffic to the trash-to-steam facility away from neighborhood streets and to monitor air quality throughout the city.

6.6 Sustainable chemistry course

It is against this backdrop of a community with a long history of environmental injustice and a university mission of civic engagement that we decided to develop a nonmajor course in chemical principles, using a sustainability lens and incorporating political action. CHEM 120 ("Sustainable Chemistry"), uses the current edition of the ACS textbook, *Chemistry in Context: Applying Chemistry to Society*. It is a three-credit class that meets three times a week in 50-minute periods. It fulfills the general education requirements for a science class without a laboratory and has no prerequisites beyond a basic proficiency in algebra. It is a required course for students minoring in Sustainability Science and Sustainability Management. The course goals with respect to chemistry content are similar to those for our other nonmajor classes but include specific language relating to sustainability and the environment. Table 6.1 maps these learning goals to specific components of the course.

The class has now been offered twice. The grading scheme was similar in both cases: grades were based on three examinations (300 points total), seven homework assignments (100 points total), attendance and class participation (50 points), and

Table 6.1 Sustainable Chemistry learning objectives mapped to course content.

Learning goals	Course components
Students should be able to: 1. Understand the basic chemistry of the environment	Chapter 1: The Air We Breathe Chapter 3: The Chemistry of Global Climate Change Chapter 5: Water for Life (limited coverage)
2. Recognize the role of chemistry in a sustainable world	Chapter 0: Chemistry for a Sustainable Future Sustainable innovations were also discussed throughout the course in the context of each chapter
3. Understand how science explains the workings of the natural and physical world using theories and models that can be tested using experiments and observations (specifically by using chemical nomenclature and structures and representing physical phenomena mathematically)	Chapter 1.6: Classifying Mater Chapter 1.7: Atoms and Molecules Chapter 1.8: Names and Formulas Chapter 1.9: Chemical Change Chapter 2.3: Molecules and Models Chapter 3.6: Quantitative Concepts: Mass Chapter 3.7: Quantitative Concepts: Molecules and Moles
4. Use mathematical methods to solve problems	Chapter 1.1: What's in a Breath? (%) Chapter 1.2: What Else is in a Breath? (ppm) Chapter 1.14: Back to the Breath—at the Molecular Level (calculating atoms, molecules) Chapter 3.6: Quantitative Concepts: Mass Chapter 3.7: Quantitative Concepts: Molecules and Moles Chapter 4.2: Efficiency of Energy Transformation Chapter 4.3: The Chemistry of Coal
5. Interpret, make inferences, and draw conclusions from data presented in tabular or graphical form	All chapters
6. Determine if numerical results are reasonable	Same as learning goals 3 and 4
7. Examine, evaluate, and refine habits of thinking and accept ambiguity while questioning assumptions and ideas	Homework assignments Classroom discussion
8. Make claims and draw conclusions supported by the marshalling and evaluation of evidence	Homework assignments Classroom discussion
9. Synthesize divergent contents, methodologies, and models as reflective learnings and thinkers across and within disciplines	Homework assignments Classroom discussion
10. Identify environmental issues and learn to advocate at the local, state, and national levels	Environmental justice tour of Chester Advocacy Day at State Legislature in Harrisburg Field trip to Washington, DC—government and ACS Green Chemistry Institute

the Lobby Day project, detailed in Section 6.7 (100 points). The homework problems were intended to require the students to read the textbook before class. Examinations contained both quantitative and qualitative assessments of student learning.

The first class period was spent giving an overview of environmental issues and the role green chemistry plays in addressing these issues, based on Loyd Bastin's President's Lecture at Widener University in February 2015. Entitled "Chemistry for a Sustainable Future," this presentation highlighted sustainability with respect to population growth, energy consumption, and product manufacturing. The sources and finite nature of raw materials were discussed, along with waste and pollution. Concepts such as the triple bottom line, ecological footprint, cradle-to-cradle, and green chemistry were introduced as a way of thinking about environmental issues. Students were encouraged to consider how we can be more creative in developing processes that are more sustainable economically, environmentally, and socially. The theme of sustainable innovations and solutions was carried throughout the remainder of the class discussions.

About two-thirds of the regular class periods were used for chemistry content. These were loosely based on the PowerPoint slides provided by McGraw-Hill for *Chemistry in Context*, with ample time for discussion and research into local issues relating to the topic. For example, when introducing air quality (Chapter 1), students searched for current air quality values (AQIs) in Chester and talked about the $PM_{2.5}$ and ozone levels. The current levels were discussed in the context of the AQI ranges provided by the EPA and compared to other urban areas in the United States. Two class periods were used for hourly examinations with the last one (which was not comprehensive) given during the final examination period. The remaining class time was used to prepare for and participate in the civic engagement component of the course.

Given that a substantial fraction of the course was not used for chemistry content, we found that we were only able to cover about five chapters of the textbook:

Chapter 0: Chemistry for a Sustainable Future
Chapter 1: The Air We Breathe
Chapter 2: Protecting the Ozone Layer
Chapter 3: The Chemistry of Global Climate Change
Chapter 4: Energy from Combustion

In future iterations, we plan not to cover some of the content in these chapters so that we can address topics such as water quality, acid rain, and plastics.

6.7 Political and community engagement activities

To meet our goal of introducing students to environmental issues at the local, regional/state, and national levels, we incorporated three field trips into the course. At the local level, students took an environmental justice tour of Chester, led by Rev. Dr. Horace Strand of the CEP. We reserved a college van for this activity, which took students to parts of the city they were unlikely to see otherwise. Interstate 95 bisects the city, and the main campus is separated from the business district of Chester by the Avenue of the States Bridge. Dr. Harris describes in his narrative *(67)* that students and faculty historically were cautioned not to cross the bridge. In recent years, however, greater university–city collaboration in the form of a cultural program called "Bridges and Boundaries" has started to attract students, faculty and staff to downtown Chester. Nevertheless, many members of the university community do not spend time in the main part of the city.

The hour-long tour took students to historical sites, such as a plaque marking the actual location of William Penn's landing in America, a monument marking the first settlers (from Finland), and the statue of Dr. Martin Luther King, Jr., who attended Chester Theological Seminary. Students were introduced to the rich manufacturing history of Chester, passing by the sites of former shipyards and metalworks, many of which are now abandoned brownfields. They saw current businesses, such as Kimberly–Clark, Covanta, and DELCORA, and learned first-hand from Dr. Strand about the efforts of the CEP to improve the environmental quality of the city. After the tour, students were asked to reflect on their experience. The following comments were representative (bold italics are added for emphasis here and in quotes below):

> *Before the environmental tour of Chester, I was not aware of all the problems that existed and the tour opened my eyes to them. The experience was valuable to me because* **now that I'm aware of the issues I can help the community***.*

> *The most important thing to the environment in Chester in my view is the way the area views its worth (the area is regarded as trash so it is treated as such). After the tour, I realized* **how connected all the issues in Chester are***, you can't just fix one issue and not another.*

At the other end of the political spectrum, we also joined a university-sponsored bus trip to Washington, DC. For the first half of the day, students in political science, communication studies, and sustainable chemistry visited either the US Capitol or the White House in an outing organized by faculty in Political Science and Communication Studies. The first year we taught this course, students met with elected officials and their staff at the US Capitol to discuss issues of concern. The following year, we were fortunate to have a guided tour of the White House, led by David Almacy, a Widener alumnus who was a member of the communications staff of George W. Bush.

After lunch, both classes traveled to the ACS Green Chemistry Institute (GCI), where they met with a staffer and learned about GCI initiatives such as the Pharmaceutical Roundtable *(88)* and lobbying efforts to promote sustainability in the chemical and pharmaceutical industries. The students were awed by the trip to Washington, as indicated by this representative comment:

> *I thought that this experience was great for me because* **I have never done anything like this before***, and probably would never have.*

Most of the time spent in this class on civic engagement (approximately four to five class periods in total) focused on preparing for a trip to the State Capitol in Harrisburg, PA. Students were given an assignment to access the website for the state legislature *(89)* and identify pending legislation relating to sustainability issues. Each student was asked to write a short summary of the bill they thought was most interesting, and the summaries were shared with the class. After discussion, the class identified one or two bills to prepare talking points (either for or against the bill) for the legislature visit.

In one class iteration, all students wrote a position paper about the bill the class had chosen, and that was compiled into a final "white paper" by the instructor prior to the visit to Harrisburg. In another, after identifying one bill in the State House and one in the State Senate, the students drafted language for opinion postcards relating to each bill to be filled out by members of the Widener community for delivery to their individual representatives and senators. Once the language was refined, it was provided to a Communications Studies major who did the layout and added eye-catching graphics for the two postcards (Figs. 6.1 and 6.2).

Interestingly, in both class iterations, the "Growing Greener" initiative was chosen by the students *(90)*. This initiative was originally signed into law in late 1999 ("Growing Greener I") and was reauthorized in 2005 to provide funding through 2012 ("Growing Greener II"). The purpose of this initiative was to improve air and water quality (as well as quality of life) in Pennsylvania by providing funds for recreation, parks, farm preservation, acquisition of open space, restoration of watersheds, and other environmental concerns. When our classes were offered, the legislature was considering additional funding to maintain this initiative, SB705 *(91)*. The prime sponsor of this legislation, State Senator Thomas Killion, coincidentally happens to represent Chester and the surrounding area.

In preparation for the visit to the State Capitol, one class period was led by either Dr. Wes Leckrone of the Political Science Department (first iteration) or Dr. Angela Corbo of the Communication Studies and Digital Media Informatics Departments (second iteration). They instructed students on creating talking points and "elevator speeches" and worked with them to refine the messages they wanted to convey to the elected representatives and their staffs. This collaboration began years ago when Drs. Bastin, Corbo, and Leckrone served on several university committees together and worked on another interdisciplinary project.

The next step was to distribute the postcards to the campus community for signatures. Fortuitously, the university's annual Sustainability Showcase was scheduled shortly before the trip to Harrisburg. In lieu of a class meeting, students stood at a display table and handed out postcards to attendees of the showcase, while practicing their elevator speeches. Pennsylvania residents were asked to put their home address on the contact information; all others were asked to use Widener as their address. We then sorted the postcards, using zip codes, to prepare bundles for individual senators and representatives.

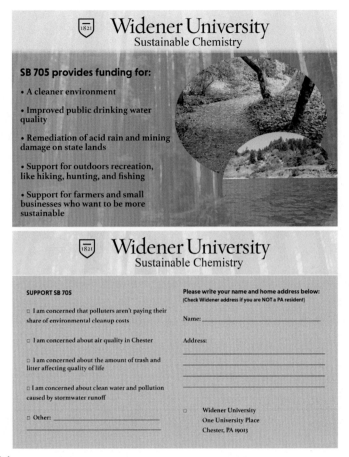

FIGURE 6.1

Top: Front side of postcard for state senators. Bottom: Reverse side of postcard for state senators.

On AICUP Lobby Day, we traveled by bus to the historic State Capitol building in Harrisburg. We were able to enter the galleries for both the House and Senate chambers to admire the architecture and furnishings. Using the staff directory, students were tasked with finding appropriate offices to deliver the postcards. Students introduced themselves to the staff and asked to speak with the representatives (Fig. 6.3). One of the chambers was in session, so we were not always able to meet with the officials directly. However, in all cases, the students received a warm welcome from the office staff and were able to deliver the postcards and their messages in person. We strongly believe that this experience gave students an understanding of how they can be active participants in the political process and advocate for positive change.

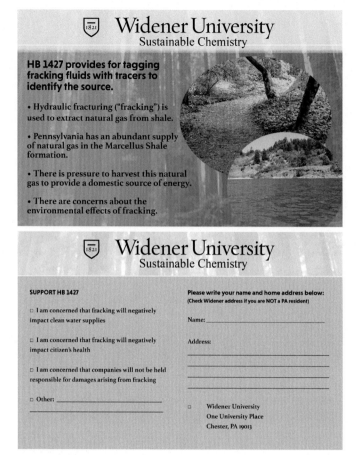

FIGURE 6.2

Top: Front side of postcard for state representatives. Bottom: Reverse side of postcard for state representatives.

As part of the assignment, the students were asked to reflect on "what you learned from the entire experience?," "what surprised you most about the Lobby Day?," and "suggestions for next year." The word cloud in Fig. 6.4 illustrates the common themes of the reflection, and the personal meetings with the representatives and senators about bills clearly resonated with the students.

The student responses were overwhelming positive and clearly demonstrated the value of incorporating the experience into the curriculum.

The class trip to Harrisburg last week was an eye opening and enjoyable experience for me. I was extremely **surprised to discover how open each representative was to talking with us and hearing our personal opinions**.

FIGURE 6.3

Widener students visiting State Representative Edward J. Neilson's office.

FIGURE 6.4

Word cloud generated from student reflections on the Lobby Day experience *(92)*.

6.8 Student reflections

At the end of the course, students were asked to reflect using a two-page essay about "what chemistry concept will you remember most from this class?"; "what sustainability concept will you remember most from this class?"; "how has your perspective of chemistry changed?"; "how has your perspective of sustainability changed?"; "what issue do you think is the most urgent to address?" and "what three recommendations would you make to address it?" The student responses (Fig. 6.5) clearly

FIGURE 6.5

Word cloud generated from student reflections on the course *(92)*.

indicated that the key concepts they took away from the course were sustainability and its relationship with chemistry and the Chester community and how their perspective on both had changed as a result of the course.

6.9 Conclusion

Civic engagement can be incorporated into a nonmajor general chemistry course by linking chemical concepts with sustainability and quality-of-life issues. Local, regional and national perspectives were highlighted with field trips and the opportunity for political advocacy. We believe this model is broadly applicable to other institutions, as environmental concerns are not confined to Chester. In circumstances where field trips to state or federal sites are not feasible, there are likely issues close to campus that could be the focal point of the civic engagement component.

6.10 Future plans

In terms of chemistry content, we feel that coverage of water quality and plastics (particularly relating to waste and recycling) is essential for our students to become more informed about sustainability. *Chemistry in Context* contains a great deal of content, so we will have to be more selective about the material we use from the earlier chapters.

For the next class iteration, we will try to schedule a visit to Harrisburg on a day other than AICUP Lobby Day. By confirming meetings with our representatives in advance and planning around the legislative session, our students will have a better opportunity for in-depth discussions with lawmakers. In addition, we hope to plan a trip to our local State Senator's office or invite him to visit with us on campus. We believe there are opportunities for greater collaboration between our students and those in Communication Studies and Political Science in terms of creating a message and communicating it effectively to legislators. It might be advantageous to try to schedule these classes at a common time.

Because students are so unaware of the history of environmental racism in Chester, we are considering showing the documentary *Laid to Waste* early in the semester. While we could use a class period for this, it might also make an excellent activity for one of our student science clubs to sponsor. Finally, through the CEP, we have made connections at local businesses, such as Covanta. A field trip to these facilities would be of interest not only to the students in our class but also to the broader campus community.

Acknowledgments

We wish to thank Dr. Angela Corbo, Communication Studies and Digital Media Informatics Departments, Widener University, for organizing the field trips to Washington, DC, and for consulting with our students on creating their messages. Dr. J. Wesley Leckrone, Political Science Department, Widener University, arranged the AICUP Lobby Day activities and consulted with our students. Students in POLS 101, SSCI 288, and CHEM 120 traveled together on our excursions. We particularly wish to thank Zunilda Jamatte, a senior Communication Studies major, for designing the postcards we delivered to state officials. The Widener University Office of Student Affairs provided grants to cover the travel expenses for our field trips. Project Pericles helped fund the course development. Finally, we also thank our tour guides: Mr. David Almacy (White House Tour), Ms. Jennifer McKellar (GCI Tour), and Rev. Dr. Horace Strand (Chester Environmental Partnership; Chester Environmental Justice Tour).

References

1. *Greater Expectations: A New Vision for Learning as a Nation Goes to College,* 2002. Washington, DC.
2. Liberal Arts Strategies for the Chemistry Classroom. In *ACS Symposium Series,* 1266, Kloepper, K. D., Crawford, G. L., Eds.; American Chemical Society: Washington, DC, 2017.
3. Stuckey, M.; H., A.; Mamlok-Naaman, R.; Eilks, I. T. The Meaning of 'Relevance' in Science Education and Its Implications for the Science Curriculum. *Stud. Sci. Educ.* **2013,** *49,* 1−34.
4. Metzger, E.; Glazier, S. Environmental Justice: Chemistry in Context for Prison Inmates and Non-Majors. In *Liberal Arts Strategies for the Chemistry Classroom*; *ACS Symposium Series*; American Chemical Society: Washington, DC, 2017, 1266; pp 167−183.

5. Tyson, J. Arsenic in Food and Water: Promoting Awareness Through Formal and Informal Learning. In *Mobilizing Chemistry Expertise to Solve Humanitarian Problems Volume 1*, Vol. 1267; American Chemical Society: Washington, DC, 2017; pp 83−97.

6. Boyer, E. L. The Scholarship of Engagement. *J. Public Service Outreach* **1996,** *1,* 11−20.

7. Ehrlich, T., Ed. *Civic Responsibility and Higher Education;* American Council on Education and the Oryx Press: Westport, CT, 2000.

8. Gallant, K.; Smale, B.; Arai, S. Civic Engagement Through Mandatory Community Service: Implications of Serious Leisure. *J. Leisure Res.* **2010,** *42,* 181−201.

9. Saltmarsh, J. The Civic Promise of Service-learning. *Liberal Educ* **2005,** *91,* 50−55.

10. Bringle, R. G.; Hatcher, J. A. Implementing Service Learning in Higher Education. *J. High. Educ.* **1996,** *67,* 221−239.

11. Howard, J. Community Service Learning in the Curriculum. In *Praxis I. A Faculty Casebook on Community Service Learning;* Howard, J., Ed.; OCSL Press, University of Michigan: Ann Arbor, MI, 1993; pp 3−12.

12. Kuh, G. D. *High-Impact Educational Practices: What They Are, Who Has Access to Them, and Why They Matter;* Association of American Colleges and Universities: Washington, DC, 2008.

13. Seifer, S. D. Service-learning: Community-campus Partnerships for Health Professions Education. *Acad. Med.* **1998,** *73,* 273−277.

14. Markus, G. B.; Howard, J. P. F.; King, D. C. Integrating Community Service and Classroom Instruction Enhances Learning: Results From an Experiment. *Educ. Eval. Policy Anal.* **1993,** *15,* 410−419.

15. Cavinato, A. G. Service Learning in Analytical Chemistry: Extending the Laboratory Into the Community. In *Active Learning*; ACS Symposium Series; American Chemical Society: Washington, DC, 2007, 970; pp 109−122.

16. Boss, J. A. The Effect of Community Service Work on the Moral Development of College Ethics Students. *J. Moral Educ.* **1994,** *23,* 183−190.

17. Bringle, R. G.; Kremer, J. F. Evaluation of an Intergenerational Service-learning Project for Undergraduates. *Educ. Gerontol.* **1993,** *19,* 407−416.

18. Cohen, J.; Kinsey, D. F. "Doing Good" and Scholarship: A Service-learning Study. *Journal. Educat.* **1994, Winter,** 4−14.

19. Giles, D. E., Jr.; Eyler, J. The Impact of a College Community Service Laboratory on Students' Personal, Social, and Cognitive Outcomes. *J. Adolesc.* **1994,** *17,* 327−339.

20. Berger, M.; Duggan, J.; Faszewski, E. E. Development of the Environmental Forum: An Integrative Approach. *Sci. Educ. Civic Engagement* **2012,** *4,* 37−42.

21. Batchelder, T. H.; Root, S. Effects of an Undergraduate Program to Integrate Academic Learning and Service: Cognitive, Prosocial Cognitive, and Identity Outcomes. *J. Adolesc.* **1994,** *17,* 341−355.

22. Eyler, J. The Civic Outcomes of Service-learning: What do we Know? *Asso. Am. Coll. Univ. Peer Rev.* **1999,** *2,* 11−13.

23. Eyler, J.; Giles, D. E., Jr. *Where's the Learning in Service-Learning? Jossey-Bass Higher and Adult Education Series;* Jossey-Bass, Inc.: San Francisco, 1999.

24. Holsapple, M. A. Service-learning and Student Diversity Outcomes: Existing Evidence and Directions for Future Research. *Michigan J. Community Serv. Learn.* **2012,** *19,* 5−18.

25. Terlecki, M.; Dunbar, D.; Nielsen, C.; McGauley, C.; Ratmansky, L.; Watterson, N. L.; Hannum, J.; Seidler, K.; Bongiorno, E.; Owens, O.; Goodman, P.; Marshall, C.; Gill, S.; Travers, K.; Jackson, J. The Crabby Creek Initiative: Building and Sustaining an

Interdisciplinary Community Partnership. *J. Community Engagement Scholarship* **2010,** *3,* 40−50.

26. Cartwright, A. Science Service Learning. *J. Chem. Educ.* **2010,** *87,* 1009−1010.
27. Hatcher-Skeers, M.; Aragon, E. Combining Active Learning With Service Learning: A Student-driven Demonstration Project. *J. Chem. Educ.* **2002,** *79,* 462−464.
28. Kalivas, J. H. A Service-learning Project Based on a Research Supportive Curriculum Format in the General Chemistry Laboratory. *J. Chem. Educ.* **2008,** *85,* 1410−1415.
29. Morgan Theall, R. A.; Bond, M. R. Incorporating Professional Service as a Component of General Chemistry Laboratory by Demonstrating Chemistry to Elementary Students. *J. Chem. Educ.* **2013,** *90,* 332−337.
30. Kuntzleman, T. S.; Baldwin, B. W. Adventures in Coaching Young Chemists. *J. Chem. Educ.* **2011,** *88,* 863−867.
31. Montgomery, B. L. Teaching the Principles of Biotechnology Transfer: A Service-learning Approach. *Electron. J. Biotechnol.* **2003,** *6,* 13−15.
32. Grover, N. Introductory Course Based on a Single Problem: Learning Nucleic Acid Biochemistry From AIDS Research. *Biochem. Mol. Biol. Educ.* **2004,** *32,* 367−372.
33. Roberts-Kirchhoff, E. S. A Service-Learning Project Focused on the Theme of National Chemistry Week: "Energy-Now and Forever" for Students in a General, Organic, and Biological Chemistry Course. In *Service Learning and Environmental Chemistry: Relevant Connections*; *ACS Symposium Series*; American Chemical Society: Washington, DC, 2014, Vol. 1177; pp 73−85.
34. Glover, S. R.; Sewry, J. D.; Bromley, C. L.; Davies-Coleman, M. T.; Hlengwa, A. The Implementation of a Service-learning Component in an Organic Chemistry Laboratory Course. *J. Chem. Educ.* **2013,** *90,* 578−583.
35. Esson, J. M.; Stevens-Truss, R.; Thomas, A. Service-learning in Introductory Chemistry: Supplementing Chemistry Curriculum in Elementary Schools. *J. Chem. Educ.* **2005,** *82,* 1168−1173.
36. LaRiviere, F. J.; Miller, L. M.; Millard, J. T. Showing the True Face of Chemistry in a Service-learning Outreach Course. *J. Chem. Educ.* **2007,** *84,* 1636−1639.
37. Kesner, L.; Eyring, E. M. Service-learning General Chemistry: Lead Paint Analyses. *J. Chem. Educ.* **1999,** *76,* 920−923.
38. O'Hara, P. B.; Sanborn, J. A.; Howard, M. Pesticides in Drinking Water: Project-based Learning Within the Introductory Chemistry Curriculum. *J. Chem. Educ.* **1999,** *76,* 1673−1677.
39. Draper, A. J. Integrating Project-based Service-learning Into an Advanced Environmental Chemistry Course. *J. Chem. Educ.* **2004,** *81,* 221−224.
40. Heider, E. C.; Valenti, D.; Long, R. L.; Garbou, A.; Rex, M.; Harper, J. K. Quantifying Sucralose in a Water-treatment Wetlands: Service-learning in the Analytical Chemistry Laboratory. *J. Chem. Educ.* **2018,** *95,* 535−542.
41. Tomasik, J. H.; LeCaptain, D.; Murphy, S.; Martin, M.; Knight, R. M.; Harke, M. A.; Burke, R.; Beck, K.; Acevedo-Polakovich, I. D. Island Explorations: Discovering Effects of Environmental Research-based lab Activities on Analytical Chemistry Students. *J. Chem. Educ.* **2014,** *91,* 1887−1894.
42. Sutheimer, S. Strategies to Simplify Service-learning Efforts in Chemistry. *J. Chem. Educ.* **2008,** *85,* 231−233.
43. Gardella, J. A.; Milillo, T. M.; Sinha, G.; Oh, G.; Manns, D. C.; Coffey, E. Linking Community Service, Learning, and Environmental Analytical Chemistry. *Anal. Chem.* **2007,** *79,* 810−818.

44. Fitch, A.; Wang, Y.; Mellican, S.; Macha, S. Peer Reviewed: Lead lab: Teaching Instrumentation With one Analyte. *Anal. Chem.* **1996,** *68,* 727A–731A.

45. Akinbo, O. T. Bottled Water Analysis: A Tool For Service-Learning and Project-Based Learning. In *Service Learning and Environmental Chemistry: Relevant Connections*; *ACS Symposium Series*; American Chemical Society, 2014, Vol. 1177; pp 149–191.

46. Burns, W. D. SENCER in Theory and Practice. In *Science Education and Civic Engagement: The SENCER Approach*; *ACS Symposium Series*; American Chemical Society, 2010, Vol. 1037; pp 1–23.

47. Sheardy, R. D.; Maguire, C. Citizens First! An Historical Perspective. In *Citizens First! Democracy, Social Responsibility and Chemistry*; *ACS Symposium Series*; American Chemical Society, 2018, Vol. 1297; pp xi–xiii.

48. Carroll, S. B. The Importance of Interface: A Tale of Two Sites. In *Science Education and Civic Engagement: The Next Level*; *ACS Symposium Series*; American Chemical Society, 2012, Vol. 1121; pp 217–242.

49. Citizens First! Democracy, Social Responsibility and Chemistry. In *ACS Symposium Series,* Vol. 1297, Maguire, C. F., Sheardy, R. D., Eds.; American Chemical Society: Washington, DC, 2018.

50. Science Education and Civic Engagement: The Next Level. In *ACS Symposium Series,* Vol. 1121, Sheardy, R. D., Burns, W. D., Eds.; American Chemical Society: Washington, DC, 2012.

51. Sheardy, R. D. *Science Education and Civic Engagement: The SENCER Approach,* Vol. 1037; American Chemical Society: Washington, DC, 2010.

52. Latch, D. E.; Whitlow, W. L.; Alaimo, P. J. Incorporating an Environmental Research Project Across Three STEM Courses: A Collaboration Between Ecology, Organic Chemistry, and Analytical Chemistry Students. In *Science Education and Civic Engagement: The Next Level*; *ACS Symposium Series*; American Chemical Society: Washington, DC, 2012, Vol. 1121; pp 17–30.

53. Viviano, C. M.; Alderete, M. R.; Boarts, C.; McCarthy, M. Stop, Look, Listen: Making a Difference in the Way Future Teachers Think About Science and Teaching. In *Science Education and Civic Engagement: The Next Level*; *ACS Symposium Series*; American Chemical Society: Washington, DC, 2012, Vol. 1121; pp 111–131.

54. Eichler, J. F. Ground Level Ozone in Newton County, GA: A SENCER Model for Introductory Chemistry. In *Science Education and Civic Engagement: The SENCER Approach*; *ACS Symposium Series*; American Chemical Society, 2010, Vol. 1037; pp 109–116.

55. Railing, M. E. Introduction to Environmental Issues as a Chemistry for Non-Science Majors Course. In *Citizens First! Democracy, Social Responsibility and Chemistry*; *ACS Symposium Series*; American Chemical Society: Washington, DC, 2018, Vol. 1297; pp 43–51.

56. Odenbrett, G. C. Following the Phosphorus: The Case for Learning Chemistry Through Great Lakes Ecosystem Stewardship. In *Sustainability in the Chemistry Curriculum*; *ACS Symposium Series*; American Chemical Society: Washington, DC, 2011, Vol. 1087; pp 159–173.

57. Maguire, C.; da Rosa, J. Implementing SENCER Courses at Texas Woman's University. In *Science Education and Civic Engagement: The SENCER Approach*; *ACS Symposium Series*; American Chemical Society: Washington, DC, 2010, Vol. 1037; pp 45–61.

58. Bachofer, S. J. Studying the Redevelopment of a Superfund Site: An Integrated General Science Curriculum Paying Added Dividends. In *Science Education and Civic*

Engagement: The SENCER Approach; *ACS Symposium Series*; American Chemical Society, 2010, Vol. 1037; pp 117–133.

59. Davis, B. A.; Fisher, M. A. Science and Global Sustainability as a Course Context for Non-Science Majors. In *Sustainability in the Chemistry Curriculum*; *ACS Symposium Series*; American Chemical Society, 2011, Vol. 1087; pp 119–127.

60. Gurney, R. W.; Stafford, S. P. Integrating Green Chemistry Throughout the Undergraduate Curriculum via Civic Engagement. In *Green Chemistry Education: Changing the Course of Chemistry;* Anastas, P. T., Levy, I. J., Parent, K. E., Eds.; *ACS Symposium Series*; American Chemical Society: Washington, D.C, 2009, Vol. 1011; pp 55–77.

61. Eubanks, L. P.; Middlecamp, C. H.; Pienta, N. J.; Heltzel, C. E.; Weaver, G. C. *Chemistry in Context,* 5th ed.; McGraw-Hill: New York, 2006.

62. Newton, L. H.; Dillingham, C. K. *Watersheds: Classic Cases in Environmental Ethics;* Thomson Wadsworth: Belmont, CA, 1994.

63. Newton, L. H.; Dillingham, C. K.; Choly, J. *Watersheds 4: Ten Cases in Environmental Ethics;* Thomson Wadsworth: Belmont, CA, 2006.

64. Bouvier-Brown, N. C. Environmental Justice Through Atmospheric Chemistry. In *Service Learning and Environmental Chemistry: Relevant Connections*; *ACS Symposium Series*; American Chemical Society, 2014, 1177; pp 105–122.

65. *Widener University Fact Book 2018.* http://www.widener.edu/about/widener_leadership/administrative/institutional_research/WUFactBook2018.pdf.

66. *Widener University: Mission Statement.* http://www.widener.edu/about/vision_history/mission.aspx.

67. Harris, J. T., III The President's Role in Advancing Civic Engagement: The Widener-Chester Partnership. *New Dir. Youth Dev.* **2009,** *2009* (122), 107–126.

68. Brewer, S. E.; Nicotera, N.; Laser-Maira, J. A.; Veeh, C. Predictors of Positive Development in First-year College Students. *J. Am. Coll. Health* **2018,** 1–12.

69. May, J.R. *Sustainability at Widener University.* http://www.widener.edu/civic_engagement/sustainability/_docs/Widener%20Sustainability%20Report%202015-16.pdf.

70. Anastas, P. T.; Warner, J. C. *Green Chemistry: Theory and Practice;* Oxford University Press: New York, 1998.

71. Bastin, L. D.; Gerhart, K. A Greener Organic Chemistry Course Involving Student Input and Design. In *Green Chemistry Experiments in Undergraduate Laboratories,* Vol. 1233, Fahey, J. T., Maelia, L. E., Eds.; American Chemical Society: Washington, DC, 2016; pp 55–69.

72. *Beyond Benign Green Chemistry Commitment.* https://www.beyondbenign.org/he-green-chemistry-commitment/.

73. *Beyond Benign.* https://www.beyondbenign.org.

74. *History of Chester.* http://www.chestercity.com/about/.

75. *American Fact Finder.* https://factfinder.census.gov/faces/tableservices/jsf/pages/productview.xhtml?src=CF.

76. *American FactFinder - Selected Economic Characteristics - 2013-2017 American Community Survey 5-year Estimates - Chester City, PA.* https://factfinder.census.gov/faces/tableservices/jsf/pages/productview.xhtml?src=CF.

77. *American FactFinder - Selected Economic Characteristics - 2013-2017 American Community Survey 5-year Estimates - State of Pennsylvania.* https://factfinder.census.gov/faces/tableservices/jsf/pages/productview.xhtml?src=CF.

78. *Data USA - Chester, PA - Industries.* <https://datausa.io/profile/geo/chester-pa/#category_industries.

79. *Environmental Racism in Chester.* https://www.pubintlaw.org/cases-and-projects/chester-2/.

80. Cole, L. W.; Foster, S. R. *From the Ground up: Environmental Racism and the Rise of the Environmental Justice Movement;* New York University Press: New York, 2001.

81. *Superfund Site: Wade (ABM) - Chester, PA.* https://cumulis.epa.gov/supercpad/cursites/csitinfo.cfm?id=0301343.

82. *Covanta Delaware Valley.* https://www.covanta.com/Our-Facilities/Covanta-Delaware-Valley.

83. *DELCORA.* https://www.delcora.org.

84. *Laid to Waste: A Chester Neighborhood Fights for its Future;* Berkeley Media, LLC, 1997.

85. *City-Data - Chester, PA.* http://www.city-data.com/top2/co36.html.

86. *Health Indicators.* https://www.dvrpc.org/health/PASnapshot/pdf/3_1_Health_Asthma_Adult.pdf.

87. *Chester Environmental Partnership.* http://www.chesterenvironmentalpartnership.org/chester-environmental-partnership-home.html.

88. *Pharmaceutical Roundtable.* https://www.acs.org/content/acs/en/greenchemistry/industry-roundtables/pharmaceutical.html.

89. *Pennsylvania General Assembly.* https://www.legis.state.pa.us.

90. *What Is Growing Greener?* https://www.dep.pa.gov/Citizens/GrantsLoansRebates/Growing-Greener/Pages/What-is-Growing-Greener.aspx.

91. *Growing Greener III.* https://pagrowinggreener.org/growing-greener-iii/.

92. *WordClouds.com.* https://www.wordclouds.com.

Development and implementation of a bachelor of science degree in green chemistry

Nicholas B. Kingsley, PhD, Jessica L. Tischler, PhD

Associate Professors of Chemistry, Department of Chemistry and Biochemistry, University of Michigan-Flint, Flint, MI, United States

7.1 Introduction

With green chemistry principles becoming formally established in the 1990s, there were immediate calls for educators to disseminate these ideas to students at all levels *(1,2)*. Twenty years later, the lack of green chemistry's incorporation into the standard curriculum is still seen as a challenge to overcome *(3,4)*. Only in 2018 did the American Chemical Society's Committee on Professional Training (ACS-CPT) publish a supplemental resource to guide the adoption of green chemistry and sustainability principles into the standard chemistry curriculum *(5)*. Twenty years after the publication of *Green Chemistry: Theory and Practice*, the first textbook in this area, most program development has occurred at the graduate level with several Master's and PhD programs dedicated to green chemistry *(6,7)*. Prior to the launch of our program, we could not find evidence of a dedicated bachelor of science program in green chemistry at any institution in the United States. This is not due to a lack of faculty who are offering specific courses, laboratory experiences, degree options, or certificates *(7)*. Instead, the same obstacles that initially stalled the adoption of green chemistry into undergraduate programs were recently identified as continuing to hamper faculty. These challenges include an already packed curriculum, lack of teaching resources, and lack of colleague support within a department *(8—10)*. In this chapter, we will address how we overcame these same challenges to develop a green chemistry BS degree to meet our profession's need for graduates knowledgeable in green chemistry and our students' desire for a unique and enriching experience.

7.2 Overview of UM-Flint and its students

The University of Michigan-Flint is a primarily undergraduate institution that is part of the larger University of Michigan system. It was founded in 1956 as Flint Senior

Integrating Green and Sustainable Chemistry Principles into Education. https://doi.org/10.1016/B978-0-12-817418-0.00007-3
Copyright © 2019 Elsevier Inc. All rights reserved.
163

FIGURE 7.1

View of the University of Michigan-Flint campus from the banks of the Flint River (October 2018).

Credit: University of Michigan-Flint.

College and became part of the Michigan system of regional campuses in 1971. We became known as the riverfront campus due to our location on the banks of the Flint River in the heart of downtown Flint (Fig. 7.1). We were created to serve the residents of Flint, the surrounding counties in mid-Michigan, and the "thumb" region of Michigan. Our location results in a mix of both urban and rural students, many of whom are first-generation college students. Currently, we have over 6400 undergraduates and 1400 graduate students enrolled in over 100 undergraduate programs, 18 master's programs, and three doctoral programs *(11)*.

True to our original purpose of serving our local community, 92% of our students are Michigan residents and half are from Genesee County, the same county where we are located. Eighty percent of our alumni stay in the state post-graduation. The student body consists of 62% women and 20% underrepresented minorities *(12)*. The university has also partnered with several area school districts to create early college programs and dual-enrollment opportunities for high school students. A high proportion of these are STEM-bound students as the programs are geared toward premedical, prehealth, or preengineering studies.

The Department of Chemistry and Biochemistry is part of the College of Arts and Sciences (CAS). CAS is the largest college among the other professional units at UM-Flint. The department consistently has between 90 and 110 majors at any given time and graduates 6−13 majors per year. Our majors consist of 50% women; however, only 32% of our graduates have been women in the last five years. These numbers show that like many institutions, we have difficulties retaining majors, particularly women *(12)*. The ability of STEM disciplines to retain underrepresented minorities is also reflected in our statistics *(13−16)*. Overall, 15% of our majors are minority students, but we have only graduated two minority students in the past decade. Because of these statistics, we identified improved retention of women and minority students as a department goal in our 2015 program review for the university. Within the last seven years, 35% of our students have gone on to graduate school in the chemical sciences, 30% to professional schools, and 11% have been hired directly into industry. It is advantageous for our students that we are located between Midland and Detroit which provides employment opportunities within the chemical and manufacturing industries.

7.3 Evolution of our programs and curriculum

In the early 2000s, the degree offerings from the Chemistry Department at UM-Flint were typical for an ACS-accredited department (Table 7.1). We offered BS and BA degrees, a Teaching Certificate Program (TCP) and a minor. Most of our students selected the BS degree program, where the general, biochemistry, and environmental tracks could become ACS-certified degrees with the addition of an independent research course. These tracks shared the same foundation coursework and only differed by a few in-depth courses and electives. Despite all of these options, two-thirds of our students chose the biochemistry track.

After the retirement of several senior faculty and university pressure to grow enrollment, we reorganized to emphasize our strengths and became the Department of Chemistry and Biochemistry. We redesigned our biochemistry program and simplified our options to make the biochemistry degree become a stand-alone BS degree that was ACS certified. This helped more students find the program as catalog content moved online and was being accessed by searches for specific majors. The corresponding changes made in the 2008 ACS-CPT Guidelines allowed us to adjust the physics and math requirements that were considered a disincentive when prospective students were choosing between majoring in either biology or biochemistry.

Table 7.1 Overview of University of Michigan-Flint's chemistry degree programs in 2001.

Degrees	Programs	Additional options
Bachelor of Arts	General Program in Chemistry	
	Teacher's Certificate Program	
Bachelor of Science	General Program in Chemistry	
	Option A—Chemistry	Honors Program in Chemistry ACS Certification
	Option B—Biochemistry	Honors Program in Chemistry ACS Certification
	Option C – Environmental	Honors Program in Chemistry ACS Certification
	Option D—Materials Chemistry	Honors Program in Chemistry
Minor	Chemistry Teacher's Certificate	

During this reorganization, it was critically important to keep a full year of physical chemistry in the biochemistry program in order to maintain ACS certification. Our solution was to create a one-credit Math for Physical Chemistry course (CHM 344) that biochemistry students take concurrently with Physical Chemistry I (CHM 340). This course is taught by the same faculty member who teaches our physical chemistry lecture and can therefore select the specific math content from a full semester of multivariate calculus that is essential for success in the main lectures. The prerequisite courses for CHM 340 and CHM 344 are Calculus II and algebra-based Physics II *(17)*. This has been a good compromise between maintaining the rigor of our program, while allowing students flexibility in scheduling courses and not dissuading prospective majors who are interested in professional degrees (medical, dental, or pharmacy) postgraduation. Based on the success of this program model, we started with the biochemistry degree as a template for the green chemistry degree in order to develop a curriculum flexible enough to allow new courses.

Another significant change in our degree offerings came when we decided to end our degree option in environmental chemistry. While we considered this to still be an important subdiscipline to our profession, there were two main reasons we chose to end it. Between 2002 and 2012, the only faculty qualified to teach the courses specific to the degree option departed the university. These courses included Environmental Toxicology, Environmental Chemistry, Environmental Analytical Laboratory, Advanced Environmental Analysis Laboratory, and Environmental Physical Chemistry II. The lack of expertise among the remaining faculty coupled with a significant decline in student enrollment in these courses led to the decision to place our limited budget and personnel resources in other areas. However, an advantage of having this degree in our past was that it became easier to once again offer Environmental Toxicology in 2018, a key area in our developing a green chemistry program. As an established course in the catalog, it lessened the perceived risk of creating a new degree with many new courses critical to the program.

7.4 Moving toward green chemistry

Discussions about green chemistry at the University of Michigan-Flint began in 2010 with the hire of Dr. Nicholas Kingsley, an inorganic chemist. This now gave the department two faculty members who had backgrounds and interests in green chemistry as Dr. Jessica Tischler, our tenured organic chemist, also shared these interests. This sparked new ideas and conversations about how we could incorporate green chemistry content into existing courses and the possibility of offering a green chemistry course as an elective. However, we faced the same roadblocks that other schools report including issues related to a crowded chemistry curriculum, lack of educational resources/textbooks, and the largest impediment—lack of time.

7.4.1 Initial barriers

7.4.1.1 Faculty resources

In May 2013, Dr. Amy Cannon, the Executive Director of Beyond Benign, made a call for universities to join a new initiative, the Green Chemistry Commitment *(18)*. There was great interest in the department being one of the first signers to the Commitment; however, the department was at the start of a four year stretch of increased difficulties and roadblocks to program innovation. The first of these barriers was a lack of personnel. In May 2012, three teaching faculty retired or left the university, including our chair and the only full professor. These faculty losses left us with only two tenured and two tenure-track faculty, several new temporary faculty, and a new department chair to maintain our offerings. Requests were made to hire replacements immediately, but it became a five year process to rebuild our ranks.

7.4.1.2 Infrastructure resources

The following year brought both challenges and positive changes as we continued to rebuild our faculty. Starting in the winter of 2014, our science building underwent a major renovation. Specifically, our teaching laboratory spaces were completely redesigned and modernized while we tried to maintain our existing course offerings. This remodel dominated our department's attention until its completion in 2016.

Despite the difficulties that come with construction, we did obtain modern laboratory facilities and the opportunity to modernize our instrumentation. We purchased a 400-MHz NMR (our first high-field spectrometer) and an ICP-EOS among other key pieces of equipment that greatly expanded our research capabilities. With this new infrastructure, we finally felt confident that we had the physical resources to move forward with new initiatives such as program development.

7.4.1.3 Flint water crisis

In the fall of 2015, half of our laboratory spaces were completed and we had successfully hired a new tenure-track faculty member. It was then, seemingly overnight, Flint went from being the former General Motors town known as "Vehicle City" to being simply the headline "Flint Water Crisis" *(19,20)*. The politics of the decisions leading to the crisis has been well documented in the press *(21,22)*. The personal stories of Flint residents who continue to not have useable drinking water in 2018 have been recorded, and the faces of the ~9000 children poisoned by lead look to the country from the pages of magazines *(23)*. Common questions we are still asked are "How did this happen?" and "How are you doing?" Even today, this is a complicated question to answer when most of the city's residents do not trust the institutions meant to protect them and will not drink the water that comes from their faucet *(24)*.

Being at the epicenter of the crisis, there are a lot of ways we can answer the question of "How?" From a simplified viewpoint it came down to solubility rules:

every general chemistry student learns that lead chlorides are more soluble than lead phosphates. Although, the chemistry of municipal water systems and the prevention of lead contamination through corrosion control is well known, Flint's water had too much chloride and not enough phosphate resulting in solubilization of the lead (25–27). If there can be any positive outcomes of the Flint Water Crisis, one is that the lessons from Flint can help prevent the next crisis from happening in another city. Seeing the families and students suffer in the city where our university sits due to mistakes in chemistry compelled us to action. Although our department did not have experts in the field of water chemistry or heavy metal analysis and were not equipped to assist in the analysis of thousands of samples, we were experts in teaching. We realized we wanted our students and alumni to practice chemistry with a conscience; to be aware of the social and environmental consequences of their decisions in the laboratory and throughout their career. It was this desire to broaden our department's curriculum that lead us back to green chemistry.

7.4.2 Departmental review, goals, and opportunities

This period of faculty turnover left us short-staffed, while the construction and water crisis added to our stress, but it also provided us with a fairly young group of faculty open to new curriculum ideas and inspired to make changes. Also at this time, we completed a university-level program review that provided the opportunity to reflect on the previous program changes and create a new five year plan. Two goals came out of this review. One was to focus on the retention of our majors from freshman year to graduation, particularly women and underrepresented minorities who we lost at a much higher rate than white, male students. Second, we all agreed to develop something in the area of green chemistry even if it was just a dedicated elective course. Being in the midst of the water crisis, these two goals felt connected. We hoped that the coupling of social and environmental justice awareness to the innovative solutions associated with green chemistry would be appealing to a wider set of prospective students. Other schools have found greater student engagement to be an outcome of their greener curriculum (28–30).

It was within this context that another opportunity presented itself. We were approached by our School of Health Professions (now College of Health Sciences) to see if we could begin offering a toxicology course again. They wanted to restart their degree in environmental health sciences, and their national certification required a course in toxicology. Previously our environmental toxicology had fulfilled this requirement. Serendipitously, we were hoping to obtain a new faculty post in biochemistry. We decided to modify our post-request the following year to require biochemistry candidates to have secondary expertise in toxicology to allow this cross-unit collaboration. Although the applicant pool was smaller than our previous searches due to this extra requirement, we found an excellent candidate. They started in 2017 and were able to also establish a research program in toxicology. This hire turned out to be key in our creation of a green chemistry curriculum.

7.4.3 Finding support networks

The fall of 2016 brought finality to many ongoing changes in the department allowing it to move forward from the prior delays and struggles. The building renovation was complete, a second tenure track faculty in organic chemistry was hired, and two junior faculty were awarded promotion to associate professor with tenure. These events felt like an inflection point in the direction of the department. This stabilization allowed the department to further develop goals for the future and move forward with confidence.

7.4.3.1 Support in the green chemistry community

At this time, Dr. Kingsley attended the 24th Biennial Conference on Chemistry Education (BCCE) and the Michigan Green Chemistry and Engineering conference. The goal of attending these conferences was to engage in conversations around current green chemistry practices and see how the department could progress in this area. This mindset resulted in attending the Toxicology for Chemists Symposium and Workshop coordinated by the Molecular Design Research Network (MoDRN) in collaboration with Beyond Benign at the 2016 BCCE. The symposium was focused on faculty members who were integrating toxicology into their courses or programs while the workshop provided curriculum resources for use in chemistry classrooms. These sessions were a landmark moment for our program development. It was inspirational to see how green chemistry had been a transformative educational piece for many programs and that we were already on the right track with the decision to search for a new faculty member with toxicology experience.

The Michigan Green Chemistry and Engineering Conference occurred in November 2016 and brought further validation of our ideas around green chemistry initiatives. This conference brought together people from academia and industry across Michigan to share information centered on green chemistry and engineering. It was surprising that a vast majority of attendees were not from higher education but rather from industry. This was surprising as many of the discussions and presentations at the conference were specifically focused on engaging higher education to move green chemistry forward. It was clear from this conference that industries in Michigan and surrounding areas were excited about the notion of academic institutions taking a leading role in supporting and promoting the importance of green chemistry.

Reflection at this point brought us back to Beyond Benign. We were lucky to gain support within our department and the Dean's office to solidify our role in the world of green chemistry education. We decided the best place to start was to become the 38th school to sign the Green Chemistry Commitment in February 2017 *(31)*. We were finally able to move a seven year conversation into action. By becoming signers, we were pledging to incorporate green chemistry throughout the curriculum so that all of our students regardless of course or degree would be exposed to these important concepts. As we were looking for specific ways to green our curriculum, it was proposed that expertise in our department and across campus actually allowed us to offer more in the form of a bachelor's degree in green chemistry. An initial list of courses for such a degree was developed in early 2017 consisting of one new course and many existing course offerings from across the college. The proposed degree was approved by the department in January 2017, and work commenced on our university's formal degree program approval process.

7.4.3.2 Support within the university

This timing coincided with the announcement of UM-Flint's process for developing the 2018–2023 strategic plan *(32)*. Five high-level strategic priorities were announced in early 2017. One of the priorities was "Excellent Education and Scholarship Across the Institution." The priority stated: "The University of Michigan-Flint will embody UM academic excellence by increasing support for original research, scholarship, and creative endeavors that advance knowledge and improve the quality of life for all, while also providing the highest quality of teaching, enhancing research opportunities for students, and investing in faculty and staff development" *(32)*. This priority came with a commitment to provide support and professional development for faculty to create new curricula and continue to improve teaching by employing innovative and effective methods of instruction. Extensive analysis of the strategic plan also produced six key initiatives designed to help pursue these priorities. One of these initiatives was to "develop, maintain, and revitalize high-quality and viable academic programs." It was clear that the proposed BS green chemistry degree program fit well with strategic planning processes that were occurring on campus and made it easier to gather support from across the campus and with senior level administrators.

7.4.3.3 Gathering peer feedback

At this point, two key questions remained around our proposed program: (1) without any current undergraduate programs to compare against, how confident were we that our proposed curriculum was appropriate; and (2) what was the need/demand for such a program? We took a two-tiered approach to guide us to the answers. To obtain broad input, we attended the 253rd ACS Spring National Meeting in San Francisco, CA, and the 2017 ACS Green Chemistry and Engineering Conference in Reston, Virginia. We had many conversations with members of the green chemistry education community about what we were proposing and if we were heading down the appropriate pathway. The face to face networking was also incredibly valuable to gather ideas for specific content and resources that we could incorporate into the program's curriculum.

After obtaining national feedback at these conferences, we also sought local feedback from our department's Curriculum Advisory Committee. Started in 2015, this committee was composed of external contacts from industry, faculty from area institutions, as well as our own faculty and other UM-Flint stakeholders such as the Associate Dean and the Director of Research. Some of these members were also alumni of our programs. The overall goal of this committee is to gather input on specific curricular content directly from the employers and institutions that hire and admit our students. In June 2017, our meeting was focused on the goals and coursework of our proposed Green Chemistry Program.

At this meeting, there were eight representatives from a variety of large and small companies in Michigan. There were also six faculty members from schools ranging from PUI's to PhD granting institutions. Two of the members were also alumni. After discussing the rationale for a stand-alone green chemistry degree and the specific coursework, we asked the question: "When you think of a 'green' chemist, what are the types of skills or knowledge you would expect someone to have?" It was interesting that broader skills like communication and comprehensive thinking were repeated in the responses received from the industrial members. Conversely, the academic members associated a green chemist mostly with specific green chemistry content knowledge (Table 7.2).

In our discussions, we also asked the advisory committee members what barriers our graduates from this program might face. The issue that was raised by many members was that most employers or graduate programs would likely be unfamiliar with the requirements of the degree and would assume that if green content was added, then chemistry topics were removed. Therefore, it will be critical for our graduates to be able to articulate the depth and breadth of the degree on resumes

Table 7.2 Survey of Curriculum Advisory Committee members to probe their perception of what skills or knowledge is expected from a green chemist.

Discussion question	
When you think of a green chemist, what are the types of skills or knowledge you would expect someone to have?	
Industrial Members' responses	**Academic Members' responses**
• Curious	• Research experience
• Interdisciplinary	• Need ability to pass five proficiency exams
• Ability to learn and be independent	• Would work well in interdisciplinary laboratory
• Not afraid to ask questions to know more	• Broad chemistry/scientific knowledge
• Research experience	
• Passion/drive/social skills	• Sustainability/interdisciplinarity
• Can think bigger and long term	• Critical thinking component—know process but able to consider source/toxicity/life cycle
• Social justice recognition with ability to communicate the science tying back to fundamental chemistry	• Greater awareness of sourcing, toxicity, life-cycle assessment
• Department needs more industry connection—it will help these majors	

and interviews. It would also be important that when we market the program and when the students market themselves, they present the degree as value added not knowledge lost. To further emphasize this point, we made sure that the degree will still meet the 2015 certification standards set forth by ACS-CPT *(30)*. This will be revisited in our next periodic review.

7.5 Green chemistry curriculum development

After deciding to offer a degree in green chemistry, we next had to decide between two structural options: a concentration under the bachelor's degree in chemistry or a stand-alone bachelor's degree in green chemistry. Through discussions with our Curriculum Advisory Committee and within the department's faculty, the latter was chosen. There were two main factors that drove this decision: first, the switch from area options under our old degree programs to stand-alone bachelor's degrees resulted in a marked increase in biochemistry majors within the department. Second, the only concentrations under the new structure of degree programs were preprofessional track options (medical, dental, and pharmacy) that we had added in 2013 to give a choice of a non-ACS certified degree for those not pursuing a career in chemistry. It was important for us to distinguish the green chemistry degree apart from these preprofessional concentrations. These factors helped affirm our decision to offer green chemistry as a stand-alone degree.

To differentiate that the program was a unique degree, there needed to be substantive differences in the courses that were to be required. The most important component of developing the curriculum for the new program was that it must be a chemistry degree first. The emphasis of the curriculum had to be focused on chemistry while embedding the key aspects of green chemistry and sustainability throughout the course offerings. The following sections will highlight why the courses were selected for the program and its difference from the chemistry and biochemistry BS degrees.

7.5.1 Core courses

Our chemistry and biochemistry BS degrees already shared a core curriculum. Those courses include general chemistry (two semesters, lecture and laboratory), organic chemistry (two semesters, lecture and laboratory), analytical chemistry (two semesters, lecture and laboratory), physical chemistry (two semesters, lecture and laboratory), seminar (two semesters, and served as the department's capstone courses for general education), inorganic chemistry (one semester, lecture and laboratory), and chemical research (one semester). These courses were automatically included as the core curriculum in the BS green chemistry degree. Doing this gives all three BS degree programs the same foundational and in-depth chemistry content. All BS degree programs also have the same course requirements to meet general education requirements within the college *(17)*.

7.5.2 Courses specific to the Green Chemistry Program

7.5.2.1 Green learning objectives

The choices for green chemistry—specific courses were based on departmental conversations around the Green Chemistry Core Competencies from the Green Chemistry Education Roadmap and the Student Learning Objectives from Beyond Benign. The department looked for courses currently offered across the university and courses that could be developed to help address these areas.

The Green Chemistry Education Roadmap lists the following Green Chemistry Core Competencies *(33)*:

1. "Graduates will be able to design and/or select chemicals that improve product and sustainability (societal/human, environmental, and economic) performance from a life-cycle perspective.
2. Graduates will be able to design and/or select chemical processes that are highly efficient, that take advantage of alternative feedstocks, and that do so while generating the least amount of waste.
3. Graduates will understand how chemicals can be used/integrated into products to achieve the best benefit to customers while minimizing life-cycle sustainability impacts.
4. Graduates will be able to think about and make decisions taking into account life-cycle thinking and systems analysis."

Beyond Benign states the following as key Student Learning Objectives in green chemistry *(34)*. "Upon graduation, we believe all chemistry majors should have proficiency in the following essential green chemistry competencies:

1. Theory: Have a working knowledge of the 12 Principles of Green Chemistry.
2. Toxicology: Have an understanding of the principles of toxicology, the molecular mechanisms of how chemicals affect human health and the environment, and resources to identify and assess molecular hazards.
3. Laboratory Skills: Possess the ability to assess chemical products and processes and design greener alternatives when appropriate.
4. Application: Be prepared to serve society in their professional capacity as scientists and professionals through the articulation, evaluation, and employment of methods and chemicals that are benign for human health and the environment".

There is a combination of six different courses that differentiate the BS green chemistry degree from the BS in chemistry and five courses compared to the BS in biochemistry (Table 7.3). These courses are taught within the Chemistry and Biochemistry department and across campus in the Engineering and Political Science departments. Table 7.3 provides a current listing of the specific and core courses required for the green chemistry degree. Following is a complete description and rationale for each of the required courses unique to the new degree.

Table 7.3 Comparison of core and specific course requirements between our current BS degrees.

Core curriculum for all degrees	
General Chemistry I and II: Lecture and Laboratory	Organic Chemistry I and II: Lecture and Laboratory
Analytical Chemistry I and II: Lecture and Laboratory	Physical Chemistry I and II: Lecture and Laboratory
Inorganic Chemistry: Lecture and Laboratory	Calculus I and II: Lecture

Degree-specific curriculum		
Chemistry	**Biochemistry**	**Green Chemistry**
Fundamentals of Biochemistry	Biochemistry I and II: Lecture	Green Chemistry
Spectroscopy of Organic Compounds	Biochemistry I and II: Laboratory	Environmental Toxicology
Multivariate Calculus	Cell Biology	Biochemistry I and II: Lecture
Math Elective	Genetics	Math for Physical Chemistry
Two Chemistry Electives	Math for Physical Chemistry	Life-Cycle Assessment and Industrial Ecology
	One Chemistry Elective	Sustainable Design of Products and Systems
		Environmental Law and Public Policy
		One Chemistry Elective

Chemistry electives offered		
Advanced Organic	Proteomics	Green Chemistry
Polymers	Enzymology	Environmental Toxicology
Advanced Special Topics		

7.5.2.2 Green Chemistry course

This lecture course introduces and explores the 12 Principles of Green Chemistry and gives examples of applications of green chemistry in various industries. Specifically, the course covers the following: atom economy, alternative solvents, catalysis, renewable feedstocks, abiotic depletion of elements, safer design of chemicals, biodegradation, introduction to toxicology, green metrics, life-cycle analysis, risk versus hazard assessments, chemical alternatives assessment, industrial green technologies, and green washing. These topics are covered at various depths to give students an introduction to a broad range of green chemistry content. Along with these topics, students learn to analyze toxicological data and a material's properties to help choose safer chemicals. The students learn how to perform simplified chemical alternatives assessments such as Green Screen and how to determine where to find data required to do so. A capstone project for the students involves tying green chemistry to achieving the United Nations Sustainable Development Goals.

7.5.2.3 Environmental Toxicology course

This course is a carryover from the environmental chemistry BS degree that our department used to offer. In bringing toxicology back to the curriculum, we are using the title of the course as it already existed in our university catalog and was not considered a "new course" needed for the program. The faculty member teaching this course will be proposing changes in the title to more accurately reflect the topics they will be covering. The course will be entitled Fundamentals of Toxicology, and its main objective will be to provide students with a better understanding of the effects of toxicants on the body's systems. The course will begin by introducing the general principles and mechanisms of toxicology in addition to introducing the concepts of absorption, distribution, excretion of the toxicant, and toxicokinetics. The focus would then shift to the nonorgan-directed toxicity and target organ toxicity such as blood, the immune system, liver, kidney, respiratory and nervous systems, skin, and reproductive and endocrine systems. Moreover, a detailed discussion will follow, covering the various types of toxic agents such as pesticides, metals, radiation and environmental toxicology such as nanotoxicology and air pollution. The course will conclude with a consideration of the applications of toxicology by introducing subjects within the discipline such as ecotoxicology, food toxicology, analytical and forensic toxicology in order to allow students to appreciate and better understand risk assessment, and the various occupations that are responsible for carrying out these functions.

7.5.2.4 Biochemistry I and II courses

The department offers two different sequences for biochemistry lecture, a one-semester survey course of biochemistry topics and a two-semester in-depth lecture series. It was deemed important that if students were to grasp a full understanding of toxicology and the roles chemical exposure plays in human and environmental health, a more intense biochemistry sequence was needed. Biochemistry I covers structure and function of proteins, lipids, and carbohydrates, along with signal transduction. Students are expected to apply the knowledge they learn to real-life scenarios, such as case studies and explain multiple biochemical principles in writings. Biochemistry II covers metabolism of carbohydrates, lipids, proteins, and nucleic acids (both catabolism and biosynthesis). It also covers expression and transmission of genetic information. Students are expected to be able to apply these biochemical concepts to their own lives and explain some of the mechanisms behind human health and disease. The greater depth of content in this course sequence enhances a student's ability to understand important aspects of toxicology and emerging bio-based technologies often found in the application of green chemistry.

7.5.2.5 Life-Cycle Assessment and Industrial Ecology course

This engineering course is aimed at introducing basic concepts such as analytical frameworks and quantitative techniques for systematically and holistically evaluating the environmental trade-offs presented by different alternatives to enable more informed decision-making. There is specific attention paid to methodologies of life-cycle assessment and the different approaches that can be taken. Also included are discussions on the strengths and limitations of life-cycle assessment as a tool for decision-making compared to alternative approaches such as cost-benefit analysis and cost-effectiveness analysis.

7.5.2.6 Sustainable Design of Products and Systems course

This engineering course introduces students to the concepts of sustainable design of products and systems. Students look at the design of products and systems from the perspective of whole systems and life-cycle thinking for the purpose of minimizing environmental impacts. Key concepts introduced in the course are whole systems and life-cycle thinking, energy-efficient design, design for product lifetime, green materials selection, and lightweighting. This type of systems thinking is an area that is rarely included in traditional chemistry courses but was recently added to the American Chemical Society's Anchoring Concept map *(35,36)*.

7.5.2.7 Environmental Law and Public Policy course

This political science course covers political and administrative aspects of environmental regulation, including major legislation, administrative regulations, and litigation involving environmental issues. Students look at environmental policy that has been made over the last 50 years through legislation both in the Congress and in the courts. There is frequent discussion of free-rider problems, where policy does not necessarily match up with science, and the problems of laws being out of sync with the best concepts of science. This course was chosen to give students a broader perspective on environmental issues instead of just from the viewpoint of scientists. Our students who have lived through the city's water crisis can relate to environmental contamination that was caused in part from poor policy and regulation enforcement. It is important for students to understand that the design and use of safer chemicals is only part of the equation to a more sustainable future. They need to see how government policy and legislation are big factors in producing a sustainable future.

7.5.3 Mission and program approval

This combination of courses in the green chemistry degree sets it apart not only from degrees in our department but also other undergraduate programs in the country. To our knowledge it is the only undergraduate degree to offer broad, in-depth course offerings in many key areas of green chemistry as a requirement for a bachelors' degree. The green chemistry program's specific courses cover the competencies and learning objectives set forth by the Green Chemistry Education Roadmap and the Green Chemistry Commitment in an in-depth manner in full semester—long courses.

We believe this will give graduates from the program a unique opportunity to be leaders in moving green chemistry forward. It is this central belief that helped us craft the mission statement of the program, as shown below:

> *The mission of the Green Chemistry Program is to educate chemists to be ambassadors for the safer and sustainable design of chemicals and processes. Graduates will use interdisciplinary skills to responsibly shape the environmental, social, and economic impacts of chemistry.*

Our program was approved by the College of Arts and Sciences in the fall of 2017 and the university by early 2018. In the state of Michigan, all new degree programs must also undergo review from the Michigan Association of State Universities (MASU) *(37)*. MASU is a coordinating board for the 15 public universities in the state of Michigan. It provides advocacy and foster policy to maximize the collective value of the public institutions. When new programs are proposed, they are sent to each university for review. While each university handles the review in their own manner, all are given an opportunity to provide feedback or address concerns about the new program. The Green Chemistry Program was reviewed and approved by MASU in May 2018. The feedback we received was generally positive but did include a suggestion to still allow more flexibility in student electives. We are currently discussing how we could incorporate more nonchemistry electives for the next edition of the catalog as our colleagues develop interesting courses in English, sociology, economics, and business.

7.6 Conclusion

Based on the report from the First Annual Green Chemistry Commitment Summit, it is clear that barriers still exist across higher education that has prevented a broader adoption of green chemistry across the undergraduate curriculum *(8)*. At various points, we also faced many of these same challenges. We believe we have finally gathered the resources necessary to launch a full degree. Over the course of the last 17 years of faculty turnover, we have finally obtained a critical mass of people to build a sustainable program. Having at least two faculty with backgrounds in green chemistry provided a good knowledge base and the passion to push the agenda forward. Admittedly, it was also helpful that one of them was the department chair during the planning and creation process.

Being the first in anything is difficult when you are trying to create something without any examples, guidelines, or accreditation standards to follow. Having the confidence to take this leap of faith was perhaps the biggest hurdle to overcome. However, the resources and support we found from the green chemistry community of scholars have been key to our success. The insights on what should be essential elements of the program's curriculum led to the inclusion of toxicology, sustainability, life-cycle assessment, and environmental policy without compromising the core courses found in a standard chemistry degree. We know that there will be some pushback that this is a separate degree versus complete incorporation into our chemistry BS. Perhaps one day we will merge these two into one degree, but for now we believe this is the best way to get students directly out of high school interested in the program and its principles.

We currently have three senior students who will become the first graduates of the program. We asked them what made them decide to pursue green chemistry versus the traditional chemistry degree. Several of their responses are shown below:

Green chemistry seems like the future to me. The chemical industry touches all aspects of daily life, but traditional chemistry programs do not incorporate concepts such as environmental toxicology or life cycle assessment. Producing chemical products without understanding the breadth of their impacts has historically caused negative health, environmental and economic implications. The need for a safer, more sustainable chemistry is apparent and I want to be at the forefront of that change.

Student A

I decided to pursue green chemistry instead of conventional chemistry because I thought that I would be able to make a greater difference to the world one day with the specialized education that a degree in green chemistry provides.

Student B

We also asked these students if they think obtaining this degree will help in their careers. Their responses are below:

With changing governmental regulations and climate change becoming an increasingly prominent issue, I thought that getting a degree in green chemistry will be able to open more doors for me thanks to the unique and desirable skill set I have been developing over the course of my education.

Student B

There will always be interest to have an environmental consciousness while maintaining socio-economic feasibility within industry. I believe green chemistry has given me the necessary tools to approach these problems in addition to the desired chemistry.

Student C

We believe we have built a viable program that will appeal to prospective students and the prospective employers of our graduates. Our immediate focus is to increase our marketing of the degree to area schools and make sure prospective students and their teachers know what green chemistry is. Of the three majors we currently have, only one had heard of green chemistry prior to directly discussing it with us. Direct efforts by our faculty and outreach by our ACS Student Chapter, a frequent outstanding and twice Green Chapter Award winning club, to educate our student base will be needed. A huge asset in these efforts is Beyond Benign who provides resources specifically for the K-12 audience. We do realize that all curriculum reform requires assessment of its effectiveness. Currently, we are reviewing how to best assess our program's broad goals of increasing and diversifying student enrollment and graduation rate as well as student content knowledge in all aspects of green chemistry. As we look to the future, we hope that this account of how we developed the first BS degree in green chemistry will encourage other schools to increase their green chemistry offerings.

References

1. Collins, T. J. Introducing Green Chemistry in Teaching and Research. *J. Chem. Educ.* **1995,** *72,* 965−966.
2. Hjeresen, D. L.; Schutt, D. L.; Boese, J. M. Green Chemistry and Education. *J. Chem. Educ.* **2000,** *77,* 1543−1547.
3. Andraos, J.; Dicks, A. P. The State of Green Chemistry Instruction at Canadian Universities. In *Worldwide Trends in Green Chemistry Education;* Zuin, V. G., Mammino, L., Eds.; Royal Society of Chemistry: Cambridge, UK, 2015; pp 179−211.
4. Plotka-Wasylka, J.; Kurowska-Susdorf, A.; Sajid, M.; de la Guardia, M.; Namieskik, J.; Tobiszewski, M. Green Chemistry in Higher Education: State of the Art, Challenges, and Future Trends. *ChemSusChem* **2018,** *11,* 2845−2858.
5. American Chemical Society Office of Professional Training. *Green Chemistry in the Curriculum,* https://www.acs.org/content/dam/acsorg/about/governance/committees/training/acsapproved/degreeprogram/green-chemistry-in-the-curriculum-supplement.pdf.
6. Anastas, P. T.; Warner, J. C. *Green Chemistry: Theory and Practice;* Oxford University Press: New York, NY, 1998.
7. *American Chemistry Society Green Chemistry Academic Programs,* https://www.acs.org/content/acs/en/greenchemistry/students-educators/academicprograms.html.
8. Cannon, A. S. In *First Annual GCC Summit Summary Report: 25th Biennial Conference on Chemical Education, University of Notre Dame, Indiana, IN;* Beyond Benign: Wilmington, MA, 2018.
9. Eilks, I.; Rauch, F. Sustainable Development and Green Chemistry in Chemistry Education. *Chem. Educ. Res. Pract.* **2012,** *13,* 57−58.
10. Anastas, P. T.; Beach, E. S. Changing the Course of Chemistry. In *Green Chemistry Education: Changing the Course of Chemistry;* Anastas, P. T., Levy, I. J., Parent, K. E., Eds.; American Chemistry Society: Washington, DC, 2009; pp 1−18.
11. University of Michigan-Flint. https://www.umflint.edu/.
12. *University of Michigan-Flint: Common Data Set 2017−2018,* https://www.umflint.edu/sites/default/files/groups/Institutional_Analysis/documents/cds_2017-18_final_ftiac_updated_01-15-18.pdf.
13. *American Chemical Society: Diversity Statistics,* https://www.acs.org/content/dam/acsorg/membership/acs/welcoming/diversity/diversity-data.pdf.
14. The President's Council of Advisors on Science and Technology. *Engage to Excel: Producing One Million Additional College Graduates with Degrees in Science, Technology, Engineering and Mathematics,* Report to the President. https://obamawhitehouse.archives.gov/sites/default/files/microsites/ostp/pcast-engage-to-excel-v11.pdf.
15. Flynn, D. T. STEM Field Persistence: The Impact of Engagement on Postsecondary STEM Persistence for Underrepresented Minority Students. *J. Educ. Issues* **2016,** *2,* 185−214.

16. National Academy of Sciences, National Academy of Engineering, and Institute of Medicine. *Expanding Underrepresented Minority Participation: America's Science and Technology Talent at the Crossroads;* The National Academies Press: Washington, DC, 2011.

17. *University of Michigan-Flint 2018–2019 Catalog.* http://catalog.umflint.edu/.

18. Cannon, A. S. *Beyond Benign;* Personal communication: Wilmington, MA, 2013.

19. Johnson, J. *Don't Drink Flint's Water, Genesee County Leaders Warn;* M-Live.com., October 1, 2015. https://www.mlive.com/news/flint/2015/10/genesee_county_leaders_warn_do.html.

20. Davey, M. *Flint Officials Are No Longer Saying the Water Is Fine;* The New York Times, October 8, 2015; p A15.

21. Fonger, R. *Consultant Says State DEQ Gave Directives for Flint Water Treatment Plant;* MLive.com., December 10, 2018. https://www.mlive.com/news/flint/2018/12/consultant-says-state-deq-gave-directives-for-flint-water-treatment-plant.html.

22. Rosario-Ortiz, F.; Rose, J.; Speight, V.; von Gunten, U.; Schnook, J. How Do You Like Your Tap Water? *Science* **2016,** *351,* 912–914.

23. Hanna-Attisha, M. Flint Kids: Tragic, Resilient, and Exemplary. *Am. J. Public Health* **2017,** *107,* 651–652.

24. Acosta, R. *Flint Mayor Says Use Bottled Water and Filters as Pipe Replacement Continues;* MLive.com., December 4, 2018. https://www.mlive.com/news/flint/2018/12/flint_mayor_urges_residents_to.html.

25. Edwards, M.; Dudi, A. Role of Chlorine and Chloramine in Corrosion of Lead-bearing Plumbing Materials. *J. Am. Water Work. Assoc.* **2004,** *96,* 69–81.

26. Torrice, M. How Lead Ended up in Flint's Tap Water. *Chem. Eng. News* **2016,** *94,* 26–29.

27. Greenberg, M. R. Infrastructure, Environmental Justice, and Flint, Michigan. *Am. J. Public Health* **2016,** *106,* 1358–1360.

28. Gurney, R. W.; Stafford, S. P. Integrating Green Chemistry Throughout the Undergraduate Curriculum via Civic Engagement. In *Green Chemistry Education: Changing the Course of Chemistry;* Anastas, P. T., Levy, I. J., Parent, K. E., Eds.; American Chemical Society: Washington, DC, 2009; pp 55–77.

29. Ablin, L. Engaging Students With the Real World in a Green Organic Chemistry Laboratory Group Project: A Presentation and Writing Assignment in a Laboratory Class. *J. Chem. Educ.* **2018,** *95,* 817–822.

30. American Chemical Society Office of Professional Training: *ACS Guidelines and Evaluation Procedures for Bachelor's Degree Programs.* https://www.acs.org/content/dam/acsorg/about/governance/committees/training/2015-acs-guidelines-for-bachelors-degree-programs.pdf.

31. *Beyond Benign: The Green Chemistry Commitment.* https://www.beyondbenign.org/he-whos-committed/.

32. University of Michigan-Flint Strategic Plan 2017. https://www.umflint.edu/strategicplanning2017/strategic-plan-2017.

33. Holt, D. *Getting to the Core of Green Chemistry Education.* https://communities.acs.org/community/science/sustainability/green-chemistry-nexus-blog/blog/2016/05/18/getting-to-the-core-of-green-chemistry-education.

34. *Beyond Benign: The Green Chemistry Commitment Student Learning Objectives.* https://www.beyondbenign.org/he-student-learning-objectives/.

35. Holme, T. Connecting Green Chemistry Topics to the Anchoring Concepts Map. In *Presented at the Green Chemistry & Engineering Conference, Portland, OR;* June 18−20, 2018. GC&E-266.

36. MacKellar, J. Green Chemistry Education Roadmap: Systems Thinking in the Curriculum. In *Presented at the 255th ACS National Meeting and Exposition, New Orleans, LA;* March 18−22, 2018. CHED-1948.

37. *Michigan Association of State Universities.* https://www.masu.org/.

A vision for green and sustainable citizenship education at the University of Pittsburgh at Johnstown

8

Manisha Nigam, PhD

Associate Professor, Department of Chemistry, University of Pittsburgh at Johnstown, Johnstown, PA, United States

8.1 Introduction

8.1.1 About the author

I am a tenured assistant professor of organic and green chemistry at the University of Pittsburgh at Johnstown (henceforth referred to as "Pitt-Johnstown"). My academic responsibilities are as follows: (1) teach organic chemistry courses that are part of the current curriculum; (2) develop new courses that address green chemistry concepts; and (3) engage in undergraduate research in organic and green chemistry. My research is mainly pedagogical in nature, is focused on promoting the value of chemistry education in the real world, and is organized into two distinct interdisciplinary lines of research: (1) green chemistry and sustainability (in an environmental sciences context and as described in the current work) and (2) food chemistry and sustainable solutions in a global health context. Currently, I am the only faculty member in Pitt-Johnstown's Department of Chemistry who is engaged in green chemistry and sustainability research.

8.1.2 About the campus

Pitt-Johnstown is an undergraduate branch campus located approximately 80 miles from the Pittsburgh campus of the university. The current enrollment is approximately 3000 students, with an average class size of 30 students. The campus offers baccalaureate degrees in over 40 areas, along with associate degrees in the allied health fields. All faculty are required to teach at least 12 contact hours a semester. In addition to teaching, faculty are expected to engage in professional development and university/community service. Pitt-Johnstown's Department of Chemistry has six tenured/tenure-stream faculty and four faculty in the nontenure stream. Currently, the chemistry laboratories are taught by faculty who have limited background in green chemistry.

Integrating Green and Sustainable Chemistry Principles into Education. https://doi.org/10.1016/B978-0-12-817418-0.00008-5
Copyright © 2019 Elsevier Inc. All rights reserved.

8.2 Approach
8.2.1 Vision and goals

My overall vision is to raise awareness of green and sustainable principles and practices within the campus community. I am engaging students in a broad, interdisciplinary manner that will enable them to apply their education to their everyday lives on and off campus and also in their personal and professional worlds. I am also striving to leverage student efforts at promoting the long-term sustainability of the campus community. My specific goals include:

- developing undergraduate organic chemistry experiments that focus on green chemistry principles;
- designing advanced-level interdisciplinary courses that promote green and sustainability principles; and
- identifying and chartering green and sustainable projects on campus.

8.2.2 Motivation and purpose

The development of an organic/green chemistry curriculum was conceived through the guidance of Dr. Jem Spectar, President of Pitt-Johnstown. In a recent discussion, Dr. Spectar offered his perspectives on this matter.

What was your vision and motivation for creating the tenured position of Assistant Professor of Green Chemistry?

I was motivated to create this position owing to growing concerns about our environment and our planet. Results from a strategic planning survey indicated broad community interest in learning about the intersections and interdependencies among science, chemistry and sustainability.

In your assessment, what have been some of the positive outcomes to date?

Positive outcomes to date include the increase in the numbers of faculty that are promoting green chemistry research, students who are involved in research, and student presentations at the Symposium for the Promotion of Academic and Creative Enquiry (SPACE). Additional outcomes include the increase in student awareness about their environment and student proposals that promote sustainability on campus.

What can we improve upon in the short-term?

We could and should better publicize our green and sustainability programs. There is inadequate communication to prospective students about the urgency of green and sustainable principles and tying it to a sustainable planet. We could improve marketing — our current marketing channels including our website can be vastly improved in this regard.

What should we focus on in the next 5—10 years?

We should create general education courses that are interdisciplinary and focus on a multiperspective, information-based approach to understanding global problems such as climate change. General education courses will improve scientific literacy, policy decisions, and raise scientific awareness. An interdisciplinary course can connect people and perspectives from varying scientific backgrounds to formulate a holistic approach to problem-solving. Incorporation of interdisciplinary courses within the curriculum can help students and faculty understand how unification of knowledge enables effective public policy and shapes sound economic decisions. Even if people are not making the right decision, they are at least making a well-informed one!

8.3 Strengths, challenges and opportunities analysis

To realize my vision and goals, I have endeavored to leverage existing strengths and opportunities, while overcoming challenges. The following section presents a strengths, challenges and opportunities analysis of our campus.

8.3.1 Strengths

- The vision and sponsorship provided by the President of Pitt-Johnstown, who recognizes the value of green and sustainability education and initiatives
- The relatively small demographic size of the campus, which provides the ability and opportunity to easily collaborate with staff and faculty on green and sustainability initiatives
- The support of the Department of Chemistry and the Natural Science Division, who have given me the freedom of action to prototype ideas and pursue initiatives to meet my goals.

8.3.2 Challenges

- There is no systematic mechanism or process to sponsor green and sustainable initiatives on campus
- There is apathy and lack of awareness about green and sustainable practices among faculty and staff, including basic ones such as recycling
- Disseminating information across the campus population is a challenge as current mass communication mechanisms such as group emails and public service announcements are largely ignored.

8.3.3 Opportunities

- Ongoing collaborations with practitioners in the area of green chemistry education, such as Andrew Dicks and Loyd Bastin, which have enabled us to identify, design, and implement green chemistry experiments in laboratory curricula across multiple campuses
- Physical proximity of multiple universities in the Pittsburgh area, which provide networking and collaboration opportunities in green chemistry and sustainability
- Collaboration with environmentally conscious businesses such as Waste Management, which has enabled my students to learn about their green and sustainable initiatives
- Contacts with business professionals who have delivered student seminars on green and sustainable initiatives in the local region.

8.4 Development of green chemistry experiments

In this section, I describe the progress and outcomes from my first focus area, the development of undergraduate organic chemistry experiments that focus on green chemistry principles. Green chemistry is the design of chemical products and processes that reduce or eliminate the use and generation of hazardous substances *(1–5)*. Often referred to as a form of molecular-level pollution prevention, green chemistry relies on a set of 12 Principles that can be used to design or redesign molecules, materials, and chemical transformations to be safer for human health and the environment *(1)*. I have chosen to incorporate green chemistry principles into the current curriculum of sophomore-level organic chemistry experiments. Since 2010, I have leveraged student research to identify, develop, and implement environmentally friendly experiments in our chemistry laboratory curriculum, so that students are introduced to alternate methodologies for achieving chemical transformations without the use of hazardous chemicals.

8.4.1 A multiphase approach

Over the years, I have implemented a multiphase process that leverages undergraduate student research to underscore (1) the importance of microscale and green laboratories in the undergraduate chemistry curriculum and (2) associated complementary pedagogies that encourage cost saving, waste minimization, and hazard exposure limitation. Fig. 8.1 illustrates this process including the key phases, inputs, and outcomes. As depicted, it includes multiple feedback loops to enable the continuous refinement of each phase of the process, as described in the following sections.

8.4.1.1 Discovery

The discovery phase of the process is to identify appropriate experiments that my undergraduate research students will evaluate for incorporation in a laboratory setting. Since our department does not offer upper-level organic chemistry courses, I restrict my choice of experiments to those that are appropriate only for sophomore level undergraduate students. This "discovery" process includes an extensive literature survey conducted by myself which results in the identification of thousands of green procedural candidates. From these, I select appropriate ones that would be suitable for an undergraduate chemistry laboratory and viable so that they can be completed in a 3- to 4-hour practical period. Pedagogical goals and material costs are also considered.

8.4.1.2 Design and prototyping

The design and prototyping phase takes place in my research laboratory, where my undergraduate students incorporate green procedures (as described in the literature) to select experiments that have been identified in the discovery phase. Each of my research students is assigned a specific green chemistry experiment to work on

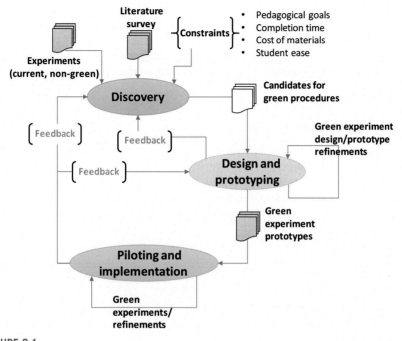

FIGURE 8.1

Multiphase process for green chemistry experiment development.

and aids in the design of the experiment. Each student may also perform a subscale prototype of the experiment design to (1) verify that the green procedure works as described; (2) fine-tune experimental variations to achieve reproducibility and robustness (i.e., the ability to withstand minor parametric variations); and (3) ensure the completion of the experiment within a 3- to 4-h time period. During prototyping, students develop and document any subprocedures or steps that provide additional directions missing from the original literature source. A key outcome of the design and prototyping phase is a detailed description of the full-scale experiment. Multiple prototyping iterations are typically required to complete this phase.

8.4.1.3 Piloting and implementation

A pilot study is undertaken in the sophomore level Organic Chemistry II laboratory (comprising 15–30 students) to test the design of a full-scale experiment in a representative classroom setting. The pilot study provides us with valuable insights on the viability of the experiment. More specifically, it enables us to address any gaps in the design of the experiment and make any modifications in order to achieve better outcomes. During the piloting phase, we account for any extra time required by students due to resource constraints (e.g., sharing of glassware/equipment). We collect data over two semesters of piloting an experiment in order to ensure consistency. The description of the experimental procedure may be revised multiple times as required. Following a successful piloting, the green chemistry experiment is ready for implementation at other collaborating campuses. The detailed experimental procedure is shared with instructors at collaborating institutions for feedback/refinement and potential integration into their laboratory curricula.

8.5 Sample green experiments

Application of the multiphase approach has resulted in the identification of several candidate experiments during the discovery phase (Table 8.1), of which three experiments reached the implementation phase. These experiments have been incorporated into the organic chemistry laboratory curriculum at Pitt-Johnstown and at other schools and are described in the following section in the context of the adopted multi-phase process.

8.5.1 A green Diels-Alder reaction

Experiment Title: "Synthesis of Substituted *N*-Phenylmaleimides and Use in a Diels-Alder Reaction: A Green Multi-Step Synthesis for an Undergraduate Organic Chemistry Laboratory." This work was recently published in the journal *Green Chemistry Letters and Reviews (6)*.

8.5.1.1 Discovery phase

In our conventional Organic Chemistry II laboratory, students had been using 2,3-dimethyl-1,3-butadiene and maleic anhydride to undertake a Diels-Alder reaction. Since the anhydride formed decomposes to a diacid (an unwanted side product), we found the percentage yield of this reaction to be lower due to the side product formation (Scheme 8.1). The purification step further reduced the yield and resulted in a longer experiment time.

   In our literature survey, we found that by using *N*-phenylmaleimide the unwanted decomposition could be avoided as the product is more stable than that derived from maleic anhydride. However, substituted *N*-phenylmaleimide derivatives are very expensive to purchase as the starting materials for Diels-Alder reactions. During our literature survey, we identified a procedure proposed by Reddy et al. *(7)* that described the green synthesis of substituted *N*-phenylmaleimides in a single step (Scheme 8.2). Unfortunately, despite repeated attempts, we were unable to reproduce those results. However, we were able to isolate and purify the substituted *N*-phenylmaleimic acid as the reaction product. We decided that this could be a viable green procedure for incorporation into the Diels-Alder experiment.

8.5.1.2 Design and prototyping phase

We designed a multistep experiment that involved the synthesis of expensive *N*-phenylmaleimide in two steps, followed by a Diels-Alder reaction in the final step.

Table 8.1 Undergraduate research projects from 2011 to 2018 (the three items highlighted in bold font have resulted in successful implementation and publication).

Year	Title of research topic
2011–12	Synthesis and Characterization of Optically Active Alcohols Using Principles of Green Chemistry
2012–13	Use of Deep Eutectic Solvents in the Perkin Reaction
2013–14	**Green Carbonyl Condensation Reactions Demonstrating Solvent and Organocatalyst Recyclability**
2012–14	**Aza-Michael Reaction for an Undergraduate Organic Chemistry Laboratory**
2013–15	Ring Opening of Cyclic Ethers to Synthesize Iodoalcohols
2014–16	One-Pot Synthesis of 6-Aminouracil Using Ionic Liquids
2015–16	Ring Opening of Cyclic Ethers to Synthesize Iodoalcohols
2016–18	**Sequential Synthesis of Substituted N-Phenylmaleimides and Their Subsequent Use in a Diels-Alder Reaction**

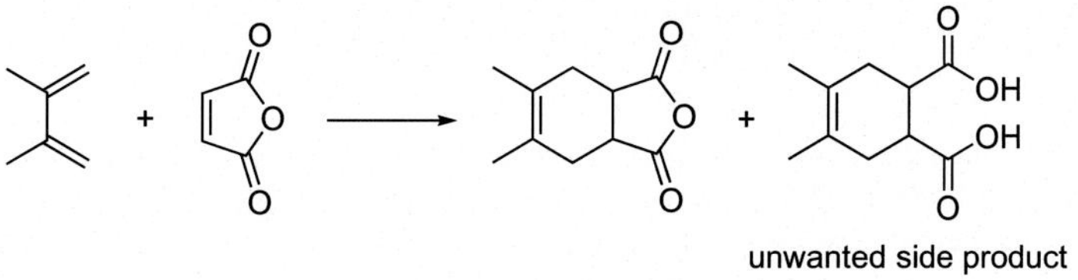

SCHEME 8.1

Diels-Alder reaction with unwanted diacid side product.

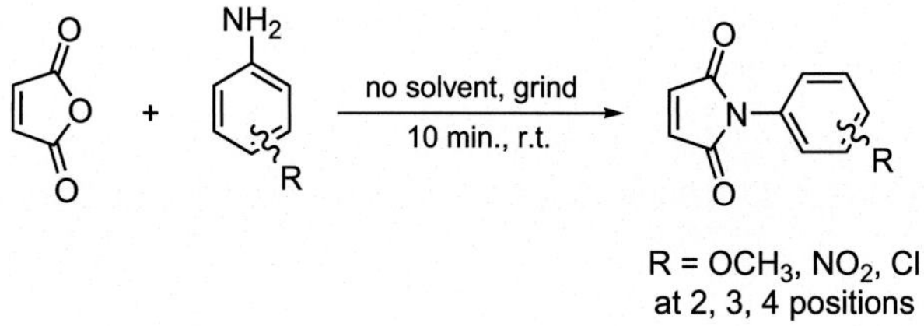

R = OCH_3, NO_2, Cl
at 2, 3, 4 positions

SCHEME 8.2

Published work that was not reproducible in our research laboratory.

Scheme 8.3 shows a green synthesis of N-(4-chlorophenyl)maleamic acid via acylation of 4-chloroaniline with maleic anhydride at room temperature.

Step 1: The amine acylation reaction in Scheme 8.3 proceeds with 100% atom economy as all the atoms from the reagents are incorporated into the maleamic

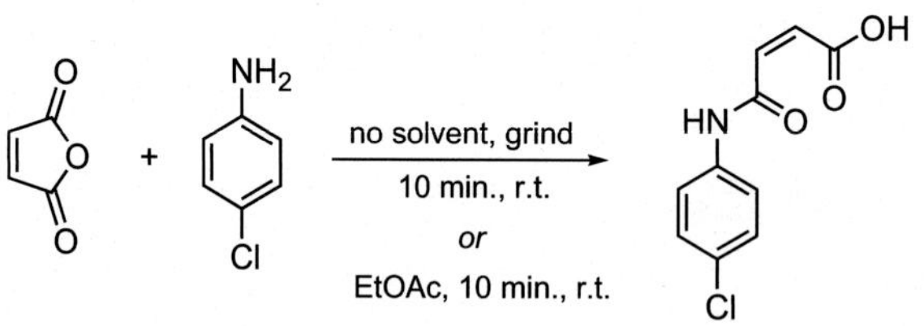

SCHEME 8.3

A green synthesis of N-(4-chlorophenyl)maleamic acid.

acid product. The synthesis was successfully accomplished under solvent-free conditions (grinding of the two solids for 10 min using a pestle and a mortar). The product is insoluble in ethyl acetate, so instead of recrystallization, trituration was performed with ∼30 mL of ethyl acetate. After trituration, the average percentage yield was 66%. An alternate acylation of 4-chloroaniline was performed by dissolving both reactants in ethyl acetate as the reaction solvent. The reaction was monitored by thin layer chromatography (TLC). The use of a solvent reduced the reaction time to 5 min, reduced the amount of ethyl acetate used (by ∼6 mL), and produced similar yields. This reaction can be analyzed using any combination of TLC, melting point, IR spectroscopy, and ^{1}H NMR spectroscopy. The IR and ^{1}H NMR spectra of the maleamic acid product are easily interpretable by students.

Step 2: This step of the multistep synthesis cyclizes N-(4-chlorophenyl)maleamic acid to N-(4-chlorophenyl)maleimide (Scheme 8.4) using a modified procedure *(8).* The average student yield was 55%. The reaction was analyzed after completion by TLC and melting point measurements. The IR and ^{1}H NMR spectra of the maleimide product are easily interpretable by students. The two-step sequence presented here provides an economic synthesis of an otherwise expensive product. The cost of the raw materials to produce 1 gram of N-(4-chlorophenyl)maleimide is $2.75, whereas the commercial cost of it from Millipore Sigma is $133 per gram.

Upon successful synthesis of the substituted N-phenylmaleimide, four dienes were then reacted with it (Scheme 8.5) and ultimately 2,5-dimethylfuran was chosen

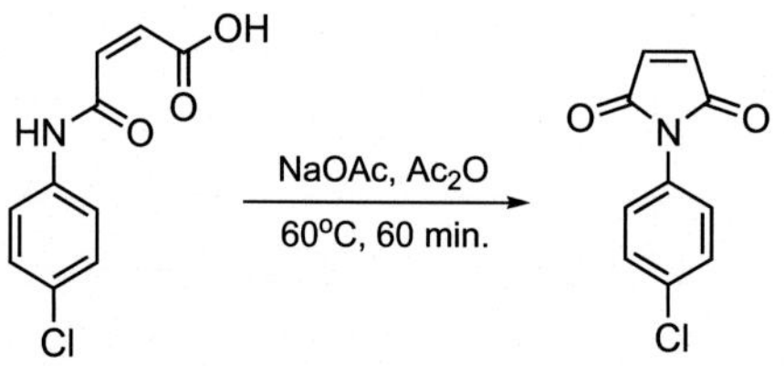

SCHEME 8.4

Synthesis of N-(4-chlorophenyl)maleimide.

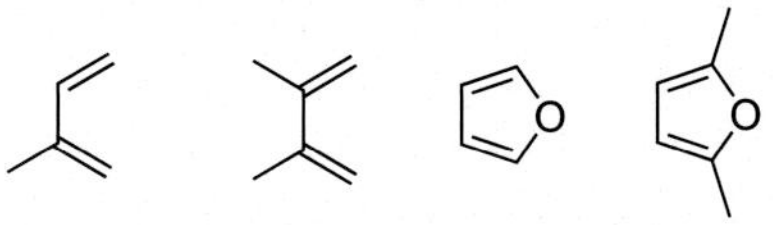

SCHEME 8.5

Different dienes reacted with *N*-(4-chlorophenyl)maleimide.

to keep the cost, physical properties, TLC monitoring, and ease of ^{1}H NMR spectral interpretation in mind.

> *Step 3:* The final step utilizes a Diels-Alder reaction between *N*-(4-chlorophenyl)maleimide and an excess of 2,5-dimethylfuran, without using any additional solvent (Scheme 8.6). After 1 h of refluxing, the reaction produces a 46% yield of *exo-N*-phenyl-7-oxabicyclo[2.2.1]-hept-5-ene-(2*S*,3*S*)-dicarboximide.

My research students repeated the experiment several times using different substituents on the *N*-phenylmaleimide benzene ring. Each step of the reaction was monitored via TLC, the intermediate compounds were purified, and their structures were confirmed via ^{1}H NMR spectroscopy and melting point measurements. The students then developed a detailed procedure in order to pilot this experiment.

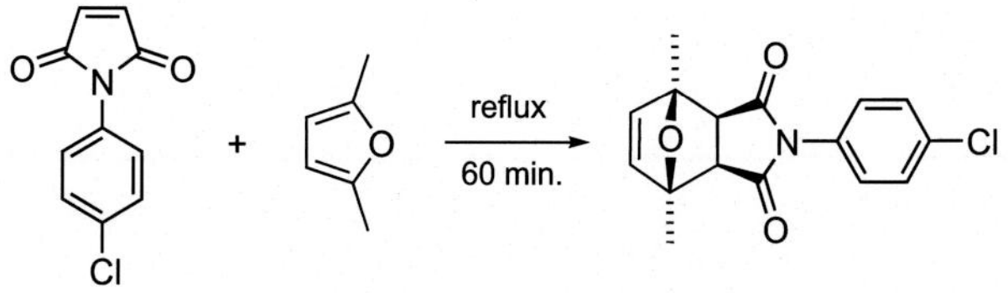

SCHEME 8.6

Diels-Alder reaction between 2,5-dimethylfuran and *N*-(4-chlorophenyl)maleimide.

8.5.1.3 Piloting and implementation phase

After two semesters of piloting this experiment at our campus, it was performed by students at Widener University where it was further modified by using a microwave reactor (which was not available at Pitt-Johnstown). This improved the versatility of the experiment due to significantly reduced reaction times. This experiment has now been fully integrated into the organic chemistry curriculum at both institutions.

8.5.2 A green aza-Michael reaction

Experiment Title: "Aza-Michael Reaction for an Undergraduate Organic Chemistry Laboratory" *(9).* This experiment addresses the issue of solvent recyclability in a sophomore organic chemistry laboratory curriculum.

8.5.2.1 Discovery phase

The Michael addition reaction is often covered after nucleophilic acyl substitutions during the second semester of sophomore organic chemistry. Most students, when asked to draw the product of a reaction between an α,β-unsaturated ester with an amine, draw an amide product as they think about the reactivity series of carboxylic acid derivatives. This presented us with an opportunity to develop an inquiry-based laboratory experiment. Based upon the IR and ^{1}H NMR spectra of the product formed, the students had to determine if the reaction proceeded to generate an amide via a 1,2-substitution mechanism or an aminoester via a 1,4-addition mechanism (Scheme 8.7).

This also presented us with an opportunity to use poly(ethylene glycol) (PEG-400) as a greener alternative solvent compared to traditional organic solvents, as it is nonvolatile, is nonflammable, exhibits low toxicity, and is stable over a variety of reaction conditions (10,11). Some other attributes of PEG-400 include ease of workup, good solvating capability, and low cost. PEG-400 is also recyclable, and minimizing waste by recycling a reaction solvent is an important principle of green chemistry. Moreover, the reaction of an α,β-unsaturated ester with an amine using traditional organic solvents (such as tetrahydrofuran or cyclohexane) is slow and does not go to completion even after 48 h (12) which is unsuitable for inclusion in a 3- to 4-h undergraduate laboratory. A literature survey indicated that the use of PEG-400 as a greener alternative to traditional organic solvents could drastically reduce reaction times to under 1.5 h (13).

8.5.2.2 Design and prototyping phase

After multiple iterations, we modified the literature procedure (13) by performing the reaction at 70°C such that the reaction time was further reduced to 45 min, and the yield was maximized (50%−100%). The experimental setup is very simple and requires basic glassware, the starting materials are inexpensive, and students are introduced to multiple green chemistry principles.

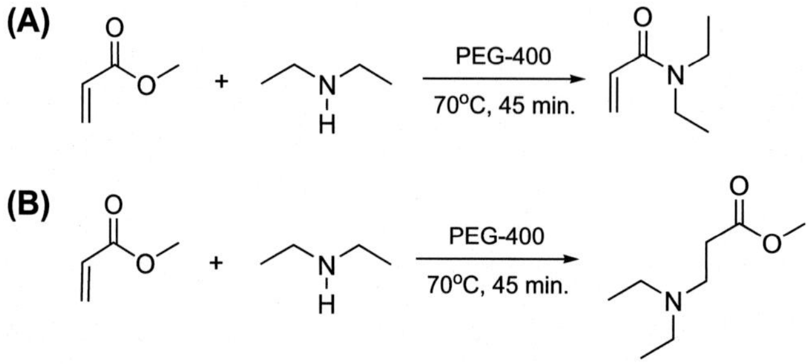

SCHEME 8.7

Two possible outcomes for the reaction between diethylamine and methyl acrylate. Pathway A is the 1,2-substitution reaction and pathway B is the 1,4-Michael addition.

8.5.2.3 Piloting and implementation phase

This experiment was piloted at Pitt-Johnstown for two semesters. The reaction was repeated several times and standardized by another undergraduate sophomore student at SUNY New Paltz during the summer of 2014. The instructor pooled the PEG-400 reaction solvent for further use resulting in a recyclability component to the experiment.

8.5.3 A green carbonyl condensation reaction

Experiment Title: "Green Carbonyl Condensation Reactions Demonstrating Solvent and Organocatalyst Recyclability" *(11)*.

8.5.3.1 Discovery phase

This experiment was based on literature procedures concerning Knoevenagel and Michael reactions in PEG-400 *(13,14)* and designed to reinforce carbonyl condensation chemistry toward the end of a second-year undergraduate course.

8.5.3.2 Design and prototyping phase

During the design phase, my research students undertook consecutive Knoevenagel and Michael reactions in one pot under green conditions to obtain a condensation product. The design phase also focused on showcasing the recyclability of both environmentally benign PEG-400 as solvent and proline as an organocatalyst (Scheme 8.8). Various other green chemistry principles were explored by the students.

8.5.3.3 Piloting and implementation phase

This experiment was piloted twice at our campus. The average student percentage yields after recrystallization in ethanol was 60%. The protocol was additionally modified at the University of Toronto and operated as an "inquiry-based" experiment in a junior level organic reaction mechanisms course. Students were not informed of the isolated product in this instance and were required to problem-solve during the laboratory to predict its structure.

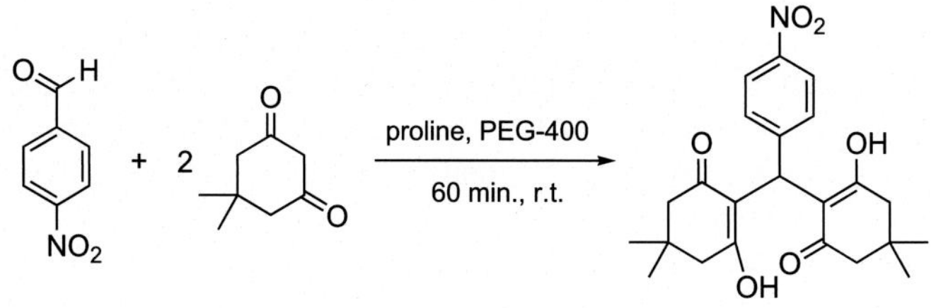

SCHEME 8.8

Consecutive Knoevenagel and Michael condensation reactions.

8.6 An advanced-level interdisciplinary course promoting green and sustainability principles

In 2012–13, I designed and implemented a new green chemistry course titled "Green Chemistry and Sustainability" for upper-level chemistry, biochemistry, and biology students who have completed the necessary prerequisite course (Organic Chemistry II). This course explores sustainability principles and the rapidly growing area of green chemistry by using an interdisciplinary approach. The factors that make green chemistry possible today and essential for the future are presented, and the effects of chemistry on the environment are investigated. The 12 Principles of Green Chemistry are studied by looking at important historical cases and current research. This course has been offered three times since Spring 2013. The course objectives are as follows:

- Introduce the concept and discipline of green chemistry and place its growth and expansion in a historical context from its birth in the early 1990's
- Demonstrate the necessity and viability of the methods of green chemistry to the chemical sciences and related disciplines
- Understand the 12 Principles of Green Chemistry as well as the tools of green chemistry including the use of alternative feedstocks or starting materials, reagents, solvents, target molecules, and catalysts
- Demonstrate how to evaluate a reaction or process and determine greener alternatives
- Discuss topics from the United States Green Chemistry Challenge Awards
- Discuss the role of the Environmental Protection Agency (EPA) in environmental laws such as the following:
  - 1970 Clean Air Act
  - 1972 Clean Water Act
  - 1972 Ocean Dumping Act
  - 1974 Safe Drinking Water Act
  - 1976 Toxic Substances Control Act
  - 1990 Pollution Prevention Act
- Examine and understand the application of interdisciplinary perspectives and innovative technologies for the development of greener routes toward improving industrial processes and products.

Business professionals were invited to the class to present green initiatives being pursued by companies such as Pittsburgh Paints and Glass (PPG). The presentation by PPG focused on the green initiatives the company is currently undertaking in manufacturing paints that are free of volatile organic compounds. The "Beyond Coal Campaign" representative for Pennsylvania (who is a member of the Sierra Club), presented a seminar on energy alternatives to coal. This was the largest campaign ever run by the 120-year-old company, which has a goal of completely transitioning from coal to cleaner energy sources by 2030. The representative also delivered a presentation on environmental hazards associated with bromides in the water supply of coal mining towns.

Waste Management (WM) was invited for several presentations to the students. These were focused on landfills that are a part of a larger network of collection, transfer, recycling, and disposal operations that are all committed to preserving the environment while providing the most comprehensive, cost-effective means of waste disposal. The students learned that landfills are designed and operated under highly regulated guidelines. Each facility employs stringent environmental controls to meet or exceed federal, state, and local regulations, and to provide maximum protection for surface and ground water.

Apart from traditional lectures, students take part in classroom discussions on environmental disasters such as Love Canal in New York, the Cuyahoga River in Ohio, the methylisocyanide accident in Bhopal, India, and the Minamata Bay accident in Japan. The students also learn about chlorofluorocarbons, DDT, and dioxins. Students have considered examples of sustainable solutions such as protecting glaciers in Peru using sawdust and mopping up oil spills with marshmallow-like macroporous gels. Specific aspects of green reactions are outlined including atom economy, green polymers, and toxicity. Some applications of green chemistry to our daily lives are discussed: these include waste treatments and the need to "recycle, reuse, and reduce." The students also deliver formal presentations of current research topics in green chemistry.

In lieu of a final examination, students develop and present project proposals for addressing sustainability on campus. They collaborate with the Physical Plant and Sodexo Dining Service personnel in developing these proposals. In their course feedback, students have indicated that they felt their proposals leverage what they learned in the classroom, and also demonstrated practical applications of their learning. Table 8.2 summarizes key project proposals developed by the students during the course, all of which were presented at the Symposium for the Promotion of Academic and Creative Enquiry (SPACE), an annual undergraduate research conference. During this event, students from every discipline on campus are given the opportunity to share the results of their class projects, independent study projects, senior projects, internships, and creative works in the form of talks and/or poster presentations.

8.7 Conceiving and executing sustainability projects at Pitt-Johnstown campus

This section describes several efforts to leverage student proposals by chartering and executing specific green/sustainability projects on campus.

Table 8.2 Student Proposals and Ideas Addressing Sustainability on Campus.

Student proposal title	Summary of proposal
Single Stream Recycling Reuse and Refuse Program	**Proposal for an easy and effective single stream recycling system on campus.**
	Proposal for the installation of bins around campus during examination week to dispose of student recyclables such as books, nonperishable food, furniture, shoes, and clothes.
Motion Sensor Bathroom Lighting	**Proposal to implement Standard Range 360° sensors with dual technology to reduce energy consumption in bathrooms of academic buildings.**
VendingMisers: A Cooler Way to a Greener Campus	**Proposal to determine the viability of installing energy-saving VendingMiser devices on vending machines on campus. A small-scale test was proposed to determine the total savings for an average campus vending machine.**
Rainwater Harvesting at Pitt-Johnstown	Proposal for the installation of a rainwater harvesting system at the engineering and science academic building.
Composting and Sustainability Awareness at Pitt-Johnstown	Proposal to enhance sustainability using the BiobiN composting system provided by the Waste Management company.
The Evolution of a Greener Campus: Converting Waste Cooking Oil to Biodiesel	Proposal for the conversion of waste oil from the dining halls to biodiesel via base-catalyzed transesterification.
EcoTraction Use on Campus	Proposal for the application of EcoTraction (a high-traction volcanic material), on campus pathways during the winter. EcoTraction works as well as road salt, but without posing the risk of corrosion to roadways, pathways, and vehicles.
Solar Water Heater on Campus	Proposal for the installation of a solar hot water heater in a student dormitory that could potentially reduce water heating bills by 50%–80%
Heat Recovery System for Engineering and Science Building Renovations	Proposal for the installation of a Vector-MD fume exhaust system in the engineering and science academic building
Laboratory Nitrile Gloves Recycling Program	Proposal to recycle nitrile gloves by collaborating with Kimberley Clark
Classroom to Classroom Recycling Education Program	Proposal to improve education on recycling efforts on campus: (1) during freshmen orientation week; and (2) during classes with the cooperation of professors who are willing to spare class time for volunteers to make a presentation

Proposals highlighted in bold font were selected for implementation as campus initiatives and are described in the following section.

8.7.1 Implementation of single stream recycling

The implementation of a campus-wide single stream recycling (SSR) program that makes recycling easy and effective for all university citizens was conceived within the context of the green chemistry and sustainability course described in Section 8.6. It was discovered that basic recycling via sorting was ineffective across the campus. The design of SSR was then driven by the student body, with support from campus administration and in partnership with WM (which operates a materials recovery facility in the local region). SSR refers to a system in which all paper fibers, plastics, metals, and other containers are mixed in a collection truck, instead of being sorted by the depositor into separate commodities (newspaper, paperboard, corrugated fiberboard, plastic, glass, etc.). After the collection, materials are separated for reuse at a materials recovery facility. The SSR implementation started in January 2015 and was well-advertised on campus. The physical plant on campus collected data regarding the quantity of SSR collected (Fig. 8.2). The total SRR tonnage collected during the time frame of January 2015–August 2016 was 23.38 tons. Implementation of the Reuse and Refuse program (discussed in the next section) attributed to a spike of SSR in the month of May as students recycled more during the campus "move-out" week.

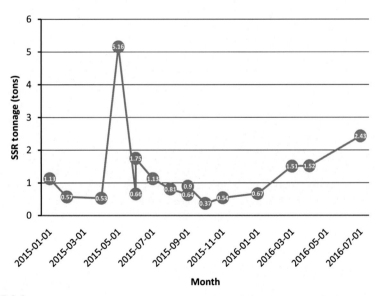

FIGURE 8.2

Tonnage report of single stream recycling (SSR) from January 2015–August 2016 (total: 23.38 tons).

8.7.2 Reuse and Refuse program

As part of the curriculum of the green chemistry and sustainability course, a second student proposal was to reduce items that are disposed of as trash by students during "move-out" week. The implementation of the Reuse and Refuse program during this week resulted in the establishment of partnerships among the Environmental Sustainability Club on campus, a local bookstore, a local youth shelter, and a local recycling plant. Students recycled qualifying objects inside large bins either outside or inside the Student Union during Examination Week. To involve the local community, a secondary location on campus was designated for the donation of their items that was collected by Goodwill Industries.

8.7.3 Installation of motion sensor lights in bathrooms

As a third example, one student enrolled in the green chemistry and sustainability course noted that lighting accounts for more than 20% of all electricity sold in the United States *(15)*. The student proposed that if the university were to invest in motion sensor lighting in the bathrooms of the Engineering and Science building, as they have done in the halls of Owen Library and Krebs Hall, they could decrease their energy costs and in return increase the amount of money to be used elsewhere. Research showed that in low or sporadically occupied areas, motion sensors could reduce lighting costs by up to 75% and could produce overall savings between 30% and 50% *(16)*. The student-proposed solution was to implement a Standard Range 360 degrees Sensor with Dual Technology manufactured by the company SensorSwitch in the campus bathrooms. These sensors not only detect very small motions such as a handwave using passive infrared detection, but they are also able to detect occupancy by using Microphonics technology to "hear" sounds that indicate a person's presence *(17)*. This would be ideal for bathrooms because they are occupied for 5—10 min at most. Prior to the implementation, the women's bathroom in the Engineering and Science building had nine fluorescent light bulbs, the men's room had five, and there was significant energy being wasted throughout the day. The student demonstrated that the benefits of the motion sensor lighting outweighed the initial investment costs, and that these costs per bathroom could be fully recovered after three years. This proposed plan was approved by the administration, and proper steps were taken to ensure a successful installation of the motion sensor provided by SensorSwitch. Fig. 8.3 shows a comparison of annual energy usage and a cost analysis *(18)*.

8.7.4 Installation of energy-saving devices on vending machines

As the final example, a fourth student proposed the viability of installing energy-saving VendingMiser devices on vending machines throughout campus. A small-scale test was undertaken to determine the total savings for the average campus

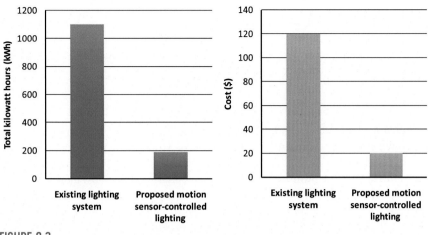

FIGURE 8.3

Comparison of the annual lighting energy usage and cost analysis.

vending machine. These machines run continuously and over the course of a year, the average one uses between 2500 and 4400 kW h of energy *(19)*. VendingMiser devices cost approximately $170 per unit *(20)* and save 46% of energy costs on average, which could result in a plan that would save both energy and money over time. To test the actual savings on campus, the student proposed the installation of a VendingMiser unit on one vending machine. In order to more accurately estimate campus-wide cost savings, the test was performed on a machine with average use as opposed to a high-traffic vending machine. By monitoring the machine's power usage, the student could compare the results both before and after the installation of the VendingMiser to experimentally determine the savings per machine. After the proposal was approved by the administration, several vending machines have been installed with the VendingMiser device. Table 8.3 shows the cost analysis of annual electricity usage. The student demonstrated that initial investment costs per vending machine could be fully recovered after 30 months.

Table 8.3 Comparison of the annual savings and cost analysis on using the VendingMiser device *(21).*

	Estimated electricity cost/year	VendingMiser savings/unit/year	Number of years to recover the cost
Low end	$178.03	$81.90	2.3
High end	$313.33	$144.13	1.3

8.8 Lessons learned

The following summarizes the lessons learned from efforts to date, along with a discussion of potential measures and actions that could be undertaken to increase the efficiency and effectiveness of future efforts.

8.8.1 Long lead times for implementing new experiments

The implementation of new green experiments into the organic chemistry laboratory curriculum has taken approximately two years per experiment. The discovery phase (which includes an extensive literature survey yielding thousands of candidate green procedures, and the identification/analysis of potentially viable ones) incurs substantial time and effort on my part. This in turn impacts my engagement in other phases (design/prototyping and piloting/implementation). During the design/prototyping phase we have often been unable to reproduce experimental results as reported in the primary literature, making us go back to the discovery phase to select a different green procedure. The reasons for this stem from a variety of factors that may substantially impact outcomes, rendering the procedure less robust. During the design process, we also must consider the cost of the chemicals to be purchased due to budget constraints. This phase may take up to two semesters ($\sim$8 months).

During piloting, we may encounter problems such as the experimental instructions being unclear. Certain aspects of an experiment that may be obvious to my research students may not be as obvious to the sophomore students performing the reaction. Time management also becomes a consideration. For example, a pre-laboratory quiz that is typically administered at the beginning of a practical period could be distributed while students are refluxing a reaction mixture for an hour. Several pragmatic adjustments must be anticipated to complete a successful pilot. Since we collect data from two semesters to ensure the achievement of consistent results, it takes one whole year ($\sim$two semesters) to complete the piloting of a single experiment. Once the piloting phase is successful, the green chemistry experiment is ready for implementation at other collaborating university departments. This may unearth minor issues that are unique to the collaborating institution. In one instance, a minor omission of the instruction to "dry the solid product by spreading out on a watch glass" resulted in a poor percentage yield for the next step at another school, so the student instructions had to be revised.

8.8.2 Transforming student proposals to campus initiatives

The new "Green Chemistry and Sustainability" course has yielded several student proposals for campus initiatives, which were initially presented at SPACE. Students worked with the physical plant and campus administration staff to analyze the viability of a few select projects. I chose specific ideas that seemed to be suitable for implementation on the campus. However, some of the student proposals were deemed to be impractical for western Pennsylvania. Either the sustainable project would require very high upfront costs to implement or the proposals would not deliver tangible benefits within the region. For instance, due to the substantial precipitation in the surrounding area, a rainwater harvesting proposal was not deemed to be practical. Other ideas were simply not economically viable owing to the campus budgetary priorities and allocations. In addition to seeking alternative funds (e.g., grants from the main University of Pittsburgh campus), I am also attempting to understand the budgetary cycles and processes for my students to factor into their proposals.

8.8.3 Cultural and regulatory challenges with recycling

The SSR program, which was implemented in 2015 with strong support from the campus administration leadership, has continued to face cultural challenges. Rural western Pennsylvania is not as environmentally conscious as the major cities. There is uneven implementation of green and sustainable practices by local governments across the region. Many citizens of this region have not been observing systematic recycling practices in their daily lives, and when they come to the campus (as a student, faculty, or staff member), they bring their old habits. Although SSR bins were clearly marked with instructions, and the program was well marketed, the SSR tonnage never trended upward as expected. This indicated that campus citizens were not using the SSR bins and/or they were using them improperly. For instance, placing food waste in an SSR bin would result in all the contents being treated as trash, and not as recyclable items.

In 2018, the SSR program faced regulatory challenges as described by the following campus-wide announcement issued by the administration: "In 2018, Pitt-Johnstown restructured the campus-wide recycling program to adhere to changes in the worldwide market regulations for recycling. Due to this, single stream recycling is no longer being implemented on campus. New signs with complete instructions and thoughtful placement of containers are in place to alleviate contamination while recycling. Recycling in residence halls has also changed, and Student Affairs has already begun to educate our students on the importance of mindfully recycling to avoid contamination." Although the recent regulatory challenges were unfortunate and beyond the control of the administration, the mindsets of the campus citizenry toward recycling could and should be actively monitored (e.g., via surveys, recycling tonnage reports, etc.) in order to better inform recycling marketing/communication initiatives aimed at changing existing attitudes across the campus.

8.8.4 Research experiences for undergraduate students

Addressing green chemistry concepts in the context of education has enabled me to encourage student research, engagement, and education on the topic of green chemistry and sustainability, and also to enhance student presentations skills. Efforts described in this chapter have provided ample research experience for undergraduate students enrolled in the courses I teach and/or engaged in my research group. Benefits include a better understanding of published literature, a balance between independence and collaboration, and efficiency in their learning experience. To emphasize this, I present below some select testimonials from my current research students and alumni who are pursuing preprofessional careers and/or postgraduate research.

From an alumnus, who is currently a first-year medical student at Temple University:

As a medical student, I am always involved with research in some way, whether it's reading articles, designing studies, or performing the actual research. My undergraduate research experience benefitted me the most by allowing me to better critique articles and understand or question why researchers chose the methods they did. Determining if an article is significant and applicable to ones' practice is a fundamental characteristic all physicians need. In addition, a valuable lesson I learned was to appreciate the time and dedication it took to get where we are with our vast knowledge that we now have.

From an alumnus, who is currently an associate scientist in the pharmaceutical biochemistry method development group at Eurofins Lancaster Laboratories:

My two years as a research student at Pitt-Johnstown helped mold me into the scientist I am today. I learned how to critically read scientific papers and other research studies to aid us with the knowledge to provide the highest success in the research we were conducting ourselves. From here, we devised a procedure and took it into the lab where I learned never to become discouraged and to learn from failures. These key aspects are still beneficial to me now that I am working in a pharmaceutical analysis lab. I must critically read our client methods to have our testing go correctly. If failures occur, I can look at the problems and troubleshooting what needs to be reevaluated to ensure a successful test the next time around.

From a sophomore student who is majoring in chemical engineering, with a goal of working in the chemical industry after graduation:

Research was a very different experience than what I had anticipated. My first thoughts were about all the new discoveries I would make and how I could use my knowledge from organic chemistry to formulate new substances that would have intriguing and unique properties. However, I was not aware of the difficulties and problems that lay ahead. One of the lessons that I learned quickly was how long and repetitive research really was. It took months to just work out a procedure for the molecule I was synthesizing. As of this writing, the procedure is still being perfected and finalized. Despite the repetition, it made me extremely efficient in the lab such that I can perform multiple tasks at once. Altogether, this research experience has improved my critical thinking skills and increased my overall efficiency.

From a junior student, who is majoring in biology and wants to pursue research in the field of public health:

Participating in research as an undergraduate student has showed me my career path. I was able to learn the etiquette of laboratories used for research purposes. More importantly, I had the opportunities to meet other students and professors who did similar work and gain their wisdom to use in my future career. It solidified my interest in being a part of the scientific advancements that will change the lives of the people in the future.

From a junior student, who is majoring in biology and wishes to pursue a career in medicine:

My experience with research has been very rewarding. Each failed experiment makes me think about what I can do to improve my research methodology and about possible reasons why the experiment failed. Research experience has affected my education by making me time efficient. I was told once by a professor to think about all the experiments that I would be performing that particular day and then do the setup for each part while I wait.

Acknowledgments

I would like to acknowledge the Pitt-Johnstown administration and the Department of Chemistry whose support was essential to my green chemistry research. I would also like to thank my research collaborators: Andrew Dicks, Loyd Bastin, Preeti Dhar, and Sam Martinus. Finally, I would like to express my gratitude to my undergraduate research students who tirelessly work in my laboratory on various projects.

References

1. Anastas, P. T.; Warner, J. C. *Green Chemistry: Theory and Practice;* Oxford University Press: New York, NY, 1998.
2. Matlack, A. S. *Introduction to Green Chemistry;* Marcel Dekker: New York, NY, 2001.
3. Lancaster, M. *Green Chemistry: An Introductory Text,* 3rd ed.; Royal Society of Chemistry: Cambridge, UK, 2016.
4. Hjeresen, D. L.; Schutt, D. L.; Boese, J. M. Green Chemistry in Education. *J. Chem. Educ.* **2000,** *77,* 1543−1544.
5. Collins, T. J. Introducing Green Chemistry in Teaching and Research. *J. Chem. Educ.* **1995,** *72,* 965−966.
6. Bastin, L. D.; Nigam, M.; Martinus, S. J.; Maloney, J. E.; Benyack, L. L.; Gainer, B. Synthesis of Substituted *N*-phenylmaleimides and use in a Diels-Alder Reaction: A Green Multi-step Synthesis for an Undergraduate Organic Chemistry Laboratory. *Green Chem. Lett. Rev.* **2019,** *12,* 127−135.

7. Reddy, Y. D.; Reddy, C. V. R.; Dubey, P. K. Green Synthesis of *N*-substituted Imides. *Afro Asian J. Sci. Tech.* **2014,** *1,* 5–9.

8. Searle, N. E. *Synthesis of N-Aryl-Maleimides,* 1948. US 2,444,536.

9. Nigam, M.; Rush, B.; Patel, J.; Castillo, R.; Dhar, P. Aza-Michael Reaction for an Undergraduate Organic Chemistry Laboratory. *J. Chem. Educ.* **2016,** *93,* 753–756.

10. McKenzie, L. C.; Huffman, L. M.; Hutchison, J. E.; Rogers, C. E.; Goodwin, T. E.; Spessard, G. O. *J. Chem. Educ.* **2009,** *86,* 488–493.

11. Stacey, J. M.; Dicks, A. P.; Goodwin, A. A.; Rush, B. M.; Nigam, M. Green Carbonyl Condensation Reactions Demonstrating Solvent and Organocatalyst Recyclability. *J. Chem. Educ.* **2013,** *90,* 1067–1070.

12. Byrd, K. M. Diastereoselective and Enantioselective Conjugate Addition Reactions Utilizing α,β-unsaturated Amides and Lactams. *Beilstein J. Org. Chem.* **2015,** *11,* 530–562.

13. Kumar, D.; Patel, G.; Mishra, B. G.; Varma, R. S. Eco-friendly Polyethylene Glycol Promoted Michael Addition Reaction of α,β-unsaturated Carbonyl Compounds. *Tetrahedron Lett.* **2008,** *49,* 6974–6976.

14. Liu, Y.; Liang, J.; Liu, X. H.; Fan, J. C.; Shang, Z. C. Polyethylene Glycol (PEG) as a Benign Solvent for Knoevenagel Condensation. *Chin. Chem. Lett.* **2008,** *19,* 1043–1046.

15. *Frequently Asked Questions Information on Compact Fluorescent Light Bulbs (CFLs) and Mercury.* https://www.energystar.gov/ia/partners/promotions/change_light/downloads/Fact_Sheet_Mercury.pdf.

16. VonNeida, B.; Maniccia, D.; Tweed, A. *An Analysis of the Energy and Cost Savings Potential of Occupancy Sensors for Commercial Lighting Systems.* https://www.lrc.rpi.edu/resources/pdf/dorene1.pdf.

17. *CMR PDT 9: Standard Range 360° Sensor.* https://www.acuitybrands.com/products/detail/757191/Sensor-Switch/CMR-Series/Ceiling-Mount-Line-Voltage-Sensors.

18. Bluejay, M. *How Much Electricity Costs, and How They Charge You.* https://michaelbluejay.com/electricity/cost.html.

19. *Madison Gas and Electric.* https://www.mge.com/saving-energy.

20. *VendingMiser™.* https://www.vendingmiserstore.com.

21. *Compare Penelec Electricity Rates.* www.electricrate.com/2016/01/compare-penelec-electricity-rates.

The green formula for international chemistry education

Glenn A. Hurst, PhD

Assistant Professor, Green Chemistry Centre of Excellence, Department of Chemistry,
University of York, York, United Kingdom

9.1 Introduction

After 25 years of growth, green and sustainable chemistry practices are becoming part of conventional integration throughout educational, governmental, and industrial environments across the globe. Significant impetus for this has been provided through the creation of the United Nations (UN) Sustainable Development Goals (SDGs) in 2015, which seek to address global challenges such as those relating to poverty, inequity, climate, environmental degradation, prosperity, and peace and justice *(1)*. While there has been commentary on addressing the SDGs within the context of green chemistry from research and industrial perspectives *(2)*, at the time of writing, there is little within the green chemistry educational literature as to how SDGs can be achieved through educating the next generation of chemists, chemical engineers, and policymakers at all levels including school, university, and graduate/continuing professional development. Given the interconnected nature of the SDGs, current focus within the community is implementation of strategies to facilitate a systems thinking approach into the curriculum *(3)*. Through highlighting the interdependence of components in dynamic systems, this approach lends itself to integration within green chemistry as applications of the principles of green chemistry, effective use of life-cycle analysis tools, and devising molecular design strategies all depend upon considering the reliance of reactions and processes on one another within local and global systems *(4)*.

While adoption of systems thinking within the context of green chemistry is of current interest to address the SDGs, significant work has already been completed. Upon establishing UN Agenda 21 in 1992, which outlined Education for Sustainable Development (ESD), this led to the United Nations Educational, Scientific, and Cultural Organization (UNESCO) initiating a Decade of Education for Sustainable Development from 2005 to 2014 *(5)*. Within this framework, UNESCO defined ESD as skill-oriented education to enable pupils to act responsibly today and to actively contribute toward developing their future in a sustainable way *(6,7)*. During this time and to date, a strong international commitment to green chemistry education has been demonstrated *(8–19)*. Cumulatively, this contributes toward addressing the SDGs through green chemistry education.

A coordinated implementation and associated evaluation of educational interventions within green chemistry across the world remains a challenge. It is through such international cooperation that global change may be evoked to more closely meet the SDGs *(20)*. Educational networks and professional societies provide an opportunity to collaborate through the development and application of curricular materials that can be widely utilized. Examples of such networks are G2C2 and the Green Chemistry Network *(21)*, which is an international network of professionals within green chemistry to facilitate collaboration and knowledge exchange between experts to enhance global education and outreach. The American Chemical Society (ACS) Green Chemistry Institute also has a similar remit *(22)*. Furthermore, Beyond Benign *(23)* develop and disseminate green chemistry and sustainable science educational resources through working directly with educators and a network of strategic partners across K-12 to higher education. In addition, Beyond Benign has established the Green Chemistry Commitment, a nationwide program across the United States that includes international participation, designed to encourage, empower, and celebrate entire departments of chemistry that transform their curriculum through green chemistry *(24)*. Moreover, the Network of Early-Career Sustainable Scientists and Engineers *(25)* is a global community of academic researchers and young professionals at the beginning of their careers working on or interested in solutions to today's most pressing sustainability challenges with "shaping education" being a strategic priority. Furthermore, making appropriate use of technology-enhanced learning resources with the flexibility that online learning brings to the process of teaching nationally and internationally can serve as a powerful tool to facilitate such coordinated efforts.

An alternative and complementary methodology to evoke change aligned with meeting the SDGs can be to introduce an impactful and transferrable template for teaching green chemistry internationally. Such an example is the master's degree course in green chemistry at the University of York, which was the first of its kind in Europe and established by Professor James Clark and colleagues in 2001 *(26,27)*. This course has been a benchmark for a number of graduate-level courses internationally with programs now available across the world in Australia, Brazil, Canada, China, Denmark, France, Greece, Hong Kong, Italy, Netherlands, Portugal, Singapore, Spain, Sweden, Switzerland, and the United States *(28)*.

9.2 The ACS Global Innovation Imperative

A recently-established scheme that has had significant global impact in the context of the SDGs is the ACS Global Innovation Imperative (Gii) *(29)*. The mission of the Gii is to create community and knowledge transfer to stimulate global scientific innovation that meets societal imperatives. This could be achieved through, for example, developing pragmatic solutions to global issues such as water quality, healthcare, or agriculture practices, to name a few. Indeed, multiple meetings across the globe in Colombia, India, Nigeria, Singapore, the United Kingdom, and the

United States have occurred to address such issues *(30−34)*. In 2016, the ACS awarded funding to facilitate a three-day workshop in Belém, Pará, Brazil, entitled "Green Chemistry Experiments for Remote Locations." Brazil was chosen to host the workshop as it is a large and diverse country with a rich biodiversity and availability of natural resources. Such features lend Brazil to be a favorable location for embedding green chemistry education in schools and universities, although there are significant challenges such as the diverse culture and educational approaches in the different regions of the country together with the lack of adequate provision of laboratories and experimental classes for teaching. Green chemistry education offers an exciting opportunity to develop a novel approach for Brazil to integrate low-cost, nontoxic chemistry experiments into Brazilian schools and universities that are relevant to the everyday lives of students, allowing them to contextualize and relate to the subject matter. Similarly, such experiments can demonstrate how students can utilize science to contribute towards building a safer and more sustainable future for their communities and the planet. Key challenges that were identified, that are summarized in a white paper with recommendations *(35)*, include the following:

- the need to develop a curriculum that enables educators to utilize activities relevant to local contexts and the knowledge of students
- the lack of a community of practice for educators in green chemistry
- educational development or coproduction (of resources) with teachers and students
- ensuring that green chemistry is integrated into the training of chemistry teachers
- improving the working environment, infrastructure, and access to materials and resources for teachers.

During the collaborative workshop between the ACS and the Brazilian Green Chemistry School, more than 30 international experts, local experts, and school teachers discussed and shared simple, low-cost, and locally relevant green chemistry experiments and activities to train the next generation of scientists in green chemistry. Other recommendations included embedding green chemistry into the national curriculum according to an integrated approach, initiating a network of green chemistry ambassadors by selecting and training highly engaged educators across Brazil, and developing online platforms to distribute transferrable educational materials. The knowledge shared and initiatives to address perceived key challenges not only have the potential to transform the landscape of green chemistry in Brazil but also serve as a transferrable model/template for use in other countries with which Brazil has similarities, especially in Latin America and the Caribbean and Africa.

9.3 **Bringing it home**

Sharing current practices during the ACS Gii meeting was enlightening, though given the challenges ahead, also provided inspiration for new educational work within green chemistry to be conducted to impact school and higher education at the internationally-leading Green Chemistry Centre of Excellence (GCCE) *(36)* at the University of York. Given the enthusiasm of undergraduate students for green chemistry within the Department of Chemistry at York together with the significant benefits of working with students as partners to enhance the curriculum *(37),* adopting a collaborative and research-led approach to develop resources for local and international implementation is part of the ethos of the Centre. Indeed, prior to the ACS Gii workshop, the GCCE established a "Sustainable Laboratories" program to integrate green chemistry and, in particular, greener research techniques throughout the curricula and laboratories *(38).* This is achieved through training laboratory technicians and Graduate Teaching Assistants as "green chemistry champions" to further embed green chemistry principles in undergraduate laboratory teaching together with working with undergraduate students to substitute hazardous and unsustainable chemicals used in their laboratory protocols. The latter project uses a Green Reagents and Sustainable Processes (GRASP) approach (Fig. 9.1) where students identify potentially hazardous chemicals within an undergraduate experiment, by, for example, consulting with the Global Harmonized System for Classification and Labeling (GHS) *(39),* the Candidate List of Substances of Very High Concern *(40)* or the European Chemicals Agency (ECHA) information via the Registration, Evaluation, Authorization, and Restriction of Chemicals (REACH) database *(41).* Following this, students then substitute such chemicals with a safe and green alternative, implementing a refined version of the activity into the undergraduate curriculum.

By employing undergraduate students as part of summer internships together with working with students through a research project component of their degree program, greener laboratory practices were incorporated. Results included solvent substitution (e.g., replacing tetrahydrofuran with the sugar-derived, 2-methyltetrahydrofuran), conducting synthetic manipulations at lower temperatures and lower stirring rates together with using reduced reagent volumes throughout experiments (which as well as being greener, has the advantage of making protocols more synthetically challenging for students, allowing for enhanced assessment

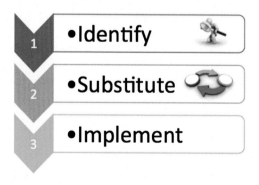

FIGURE 9.1

Green reagents and sustainable processes model.

discrimination to evaluate in-laboratory performance. Therefore, through embedding green chemistry, there is an added pedagogical value to enhance learning and associated assessment). It was exciting to see students engaging in higher order critical and creative thinking with respect to green chemistry by asking questions such as: *"Do we need to make this much product?" "What happens to it at the end?" "Do we have to heat the reaction mixture to such a high temperature?"* This work served to benefit future cohorts of students but also to those conducting the work through an invitation to give a talk at an international conference (facilitated by the students) and an invited article *(42)*. After all, future employability in the chemical sciences will require knowledge of green and sustainable chemistry: *"Manufacturers are snapping up chemists who can make their products more environmentally friendly" (43)*.

9.4 Green polymers

Following the addition of the incorporation of synthetic polymers, biological macromolecules, supramolecular aggregates and meso- or nanoscale materials into the 2015 Committee on Professional Training Guidelines in undergraduate education in chemistry *(44)*, there was significant motivation to develop new educational resources in polymer science and related areas. One such example that has long been used as part of outreach demonstrations and toys for children is silly putty (or poly(vinyl alcohol) [PVA] cross-linked with borax). This also proved to be a useful material to build laboratory experiments for undergraduates in order to help students understand rheology, which has been identified as a threshold concept *(45)*. Students can study how the kinematic viscosity varies as a function of temperature for PVA using a capillary viscometer followed by the gelation as a function of borax addition via rotational viscometry. Students may also investigate whether the resultant gel is shear thinning or shear thickening and liken the properties to ketchup or corn starch, respectively. Time-dependent effects of viscosity by varying the shear rate can even be probed together with demonstrating the Weissenberg effect, die swelling and spinnability *(46)*. Ahead of PVA gelation, the borate ion, $B(OH)_4^-$, is formed following hydrolysis of borax, $Na_2B_4O_7 \cdot 10H_2O$ (Eq. 9.1) *(47)*:

$$B(OH)_3 + 2H_2O \rightleftharpoons B(OH)_4^- + H_3O^+ \tag{9.1}$$

Students completing GRASP projects found that borax is classified as *toxic for reproduction* category 1B under Classification, Labeling, and Packaging regulations and according to the GHS, it poses a *serious health hazard* (GHS08). It is clear that such a dangerous cross-linking agent that *may damage fertility* and *may damage the unborn child* (hazard code H360) is unsuitable for educational use. As such, students participating in the project followed the GRASP methodology to utilize an alternative system. Aluminum sulfate, a coagulant for water treatment, produces a slime-like viscoelastic fluid in combination with PVA *(48)* and is suitable to demonstrate gelation, though has not been shown to form a bulk gel from which students can easily make rheology measurements. Students turned to sodium alginate, a naturally occurring anionic polymer, typically obtained from brown seaweed where, in combination with calcium chloride, gelation can occur through the formation of strong and specific ionic interactions in the form of chain-chain associations with blocks of guluronic acid residues of sodium alginate *(49)*. This system has the advantage that degelation can be instigated upon addition of trisodium citrate to the resultant gel where it is likely that citrate anions competitively combine with calcium ions, leading to a loss of chain-chain associations and subsequent disaggregation *(50)*. As such, the ability to instigate degelation has advantages regarding safe disposal of the material together with forming an additional teaching point for students, demonstrating the power of green chemistry to enhance learning in related areas. Indeed, an experiment was designed where students investigated the temperature dependence of viscosity of alginate solutions followed by gelation with calcium chloride, identification as a shear thinning gel, and disaggregation with trisodium citrate *(51)*. Owing to the cheap and benign nature of the reagents, this gelation has been performed by school students in Brazil, following development at York. Such international implementation demonstrates the accessibility and transferability of this experiment.

Formation of the calcium chloride-crosslinked sodium alginate gel provided further inspiration to develop transferrable experiments that are tailored to Brazil and similar countries, to address the Gii white paper recommendations. Brazil is the world's largest producer of oranges and uses more than 70% of the harvested fruits on the production of juices. Furthermore, the amount of processed orange is growing by approximately 10% per year, confirming the trend of the Brazilian citrus for juice production *(52)*. Given this, oranges, and more specifically, waste orange peel (WOP), is in abundance in Brazil.

Citrus peel is one of the most underutilized and most geographically diverse biowaste residues on the planet *(53)*. After citrus juice extraction, the residual peel accounts for approximately 50 wt% of the fruit, presenting significant global challenges to make use of this resource with 15.6 million metric tonnes of waste produced from 31.2 million metric tonnes of processed citrus fruit annually *(54)*. WOP is comprised of 20% dry matter (sugars, cellulose, hemicellulose, pectin, and D-limonene) and 80% water. Disposal of fresh peels is a major problem for many factories due to the pollution produced and loss of valuable resources for subsequent biorefinery processes. Furthermore, common waste treatment methodologies such as anaerobic digestion and use in animal feed are problematic due to inherently high acidity and low protein content. Consequently, due to the scale of this global issue, GRASP project students looked to develop a laboratory experiment to valorize WOP into a useable product. Further to showing students the potential to transform organic waste materials into useful products for consumer/industrial use, such an experiment would address the 2015 Committee on Professional Training Guidelines recommendations of incorporating polymeric materials into the curriculum in the context of green chemistry through adding to existing work in the literature *(46,55–58)*. Additionally, it is common in the undergraduate laboratory for students to query what will become of the product they have prepared throughout the session(s) to which a typical reply would be a description of how it will be disposed. This news is often particularly disheartening for students and sends the wrong message that sustainability is of minimal importance. Such an experiment to valorize waste is likely to have the opposite effect. Further to teaching students the role green chemistry has to play in waste valorization, this is likely to enhance student satisfaction in such laboratory experiments. Indeed, there are examples of transforming waste such as unwanted office paper (cellulose) and corncobs into useful polymeric materials such as poly(lactic acid) plastics and sunscreen respectively *(59,60)* together with using fruit and vegetable peels as adsorbents for removal of pollutants from water *(61)*. Given this background, GRASP project students developed an experiment to valorize WOP to produce a marmalade-based product that students can identify with as part of their own lives *(62)*. Students perform an aqueous extraction to form a pectin sol from WOP, which they can study the rheological properties of and in combination with a sugar solution, gelation can ensue. A gel is formed through the sugar molecules facilitating the formation of junction zones between pectin chains in close proximity to one another by promoting hydrophobic interactions between ester methyl groups and hydrogen bonding between undissociated carboxyl and secondary alcohol groups *(63)*. Following gel formation, students can subsequently study the non-Newtonian properties of the gel before having prepared a marmalade-based product. Given the widespread availability and benign nature of all reagents, this laboratory experiment has utility of use in Brazil and other countries.

9.4.1 Green and smart polymers

Further to developing transferrable laboratory experiments that may be completed within one session (e.g., 6 h), there is also the requirement to devise more advanced and exploratory research projects within green polymer chemistry where problem-based learning mini projects that reflect real-life problem-solving scenarios may find utility *(64)*. These open-ended laboratory experiments demonstrate educational value by assisting students in the development of independent practical skills through the acquisition and enhancement of research skills and critical thinking abilities *(65)*. GRASP project students identified that preparing and investigating targeted drug delivery vehicles was a very exciting area though one which had very little practical support and would lend itself to such a mini project. Smart polymeric hydrogels have been used extensively in the literature as carriers of therapeutics to a target site in the body as they are able to exhibit reversible conformational changes upon a variation in the local environment. For example, pH-sensitive hydrogels have been used to selectively deliver antibiotics to the gastrointestinal (GI) tract to treat *Helicobacter pylori* stomach infections. Cationic hydrogels such as those that are chitosan-based are suited to this purpose due to their ability to swell in acidic media. Chitosan is a natural and abundant polymer found in the exoskeletons of invertebrates, although to enhance stability, it is routinely cross-linked with agents such as glutaraldehyde, formaldehyde, or epoxy compounds, prohibiting use with undergraduate students due to serious health hazards. Similarly, temperature-sensitive poly(*N*-isopropylacrylamide)-based hydrogels have been used for drug delivery though these require the use of an ammonium persulfate initiator, which poses systemic health hazards *(66)*. As such, through the GRASP project, we embarked on identifying and substituting with a green alternative, allowing students to develop an open-ended, investigative experiment on polymers for targeted drug delivery.

Genipin is a green cross-linking agent obtained from geniposide in gardenia, an evergreen flowering plant of the coffee family, Rubiaceae. It originated in Asia and is commonly found growing wild in Vietnam, China, Korea, Taiwan, Japan, Myanmar, India, and Bangladesh. Genipin reacts with primary amines to form a conjugated and fluorescent structure with blue pigmentation *(67)*. Further to being an excellent green cross-linking agent compared to commonly-used alternatives, the fluorescent properties upon cross-linking provide a basis for students to investigate fluorescence as a function of gelation time. Usually, an additional fluorescent tag must be added to such networks in order to monitor the changes in fluorescence, while this system autofluoresces! By also incorporating PVA, this can enhance elasticity of the bulk gel owing to the hydrogen bonding with chitosan moieties.

Through the GRASP project, a laboratory experiment involving the preparation and characterization of genipin-crosslinked chitosan-PVA hydrogels was designed *(68)*. Students can monitor cross-linking via fluorescence spectroscopy, analyzing how the fluorescence intensity changes over time but also via UV-vis spectroscopy because with increasing gelation time, samples develop a dark-blue pigmentation due to oxygen-radical-induced polymerization of genipin and its reaction with amine functional groups upon exposure to air *(69)*. Students can also demonstrate the pH-sensitive swelling properties of the system through gravimetric measurements in buffered solutions of varying pH. Furthermore, students can prepare simulated GI fluid and evaluate the swelling behavior in the target environment while contrasting with a buffer of the same pH but a different ionic strength. Students can confirm that the material exhibits reversible pH-sensitive swelling behavior by periodically switching between two buffer solutions that the hydrogel is immersed in and recording the change in mass. Not only does this confirm the switchability of the material, it also shows students that the hydrogel has the potential to be coupled to a reaction that oscillates in pH to achieve pulsatile delivery of imbibed constituents. Indeed, students can also use UV-vis spectroscopy to monitor the release of a model drug compound (e.g., acetaminophen) from the network while immersed in simulated GI fluid. Finally, and rather conveniently for undergraduate use, the porous architecture can be visualized using optical microscopy. This is highly beneficial as scanning electron microscopy (SEM) is commonly used, requiring samples to be pretreated in the form of critical point drying or freeze-drying, which would alter the native hydrogel structure. SEM instrumentation is also very expensive and typically unavailable for undergraduate use. In summary, through a GRASP approach, students are now able to study complex systems in an open-ended fashion that was previously unavailable to them owing to the significant safety implications of common cross-linking agents. More introductory experiments to smart hydrogels are also available at the high school level (70,71).

9.5 Beyond polymer chemistry

Adjunct to the GRASP projects as part of the GCCE, significant recent work has been conducted to create green chemistry experiments to foster the development of independent practical skills in a number of subject areas beyond polymer chemistry to include analytical, organic, and inorganic chemistry. For example, students can assess the antibacterial properties of thyme leaf extracts through formulating and testing their own hypotheses in a series of research-led laboratory experiments, imparting knowledge and understanding of green chemistry, such as solvent selection, use of renewables, and consideration of waste streams (72). Furthermore, a highly transferrable experiment within analytical chemistry is a quantitative analysis of gasoline blends where students are involved in decision-making for a laboratory protocol examining the use of green dyes for assessing bio-ethanol-gasoline blends (73). This experiment has been implemented with Brazilian students and is both timely and relevant as gasoline has been commercialized in Brazil since 1931 and contains approximately 25% bioethanol. Indeed, conversion of sugarcane to bioethanol is an established and long-standing industry in Brazil. Since 2003, cars sold in Brazil run on bioethanol-gasoline blends (flex fuel) and the first motorcycle running on flex fuel was commercially produced in Brazil in 2009. Further to the clear relevance to Brazilian students, this experiment also works well due to the availability of associated reagents and equipment.

Within inorganic chemistry, life-cycle thinking with particular regard to chemical safety and chemical waste, both in terms of human health and environmental implications has been integrated into a laboratory class (74). Students incorporate a section on sustainability into their lab reports and devise their own modifications to enhance the sustainability of the original laboratory procedure based on safety information. Within organic chemistry, recent work includes structural elucidation exercises with Passerini reactions (75) and using experimental data to make direct comparisons via metrics assessment on methodologies to determine the greenest mode of amide synthesis (76). Undoubtedly, more laboratory experiments based on analytical, inorganic, materials, organic and other areas of chemistry will be developed in the coming years, though for maximal transferability, instructors should focus on relevance and availability of reagents/equipment.

9.6 Outreach in the laboratory

Further to the development of transferrable green chemistry practical experiments within higher education, there is significant demand for practical resources at the school level. In recent years, within polymer chemistry, experiments emphasizing environmentally friendly polymers with a target audience of middle and high school students have been developed to connect polymers and societal sustainability issues *(56,59,71,77,78)*. As an example, McIlrath et al. *(79)* designed an inquiry-based experiment performed in an Advanced Placement high school chemistry classroom where students polymerize ε-caprolactone to poly(ε-caprolactone), a biodegradable polyester. This work has been further developed by Knutson et al. *(58)* where students test the physical and mechanical properties of poly(ε-caprolactone) as a model for medical sutures and then design their own experiment to determine the potential for greening sutures by blending with polylactic acid. In doing so, students can study how the properties of the "suture" are varied. The activity can be adapted to be either fully open inquiry, where students develop their own questions to test, or guided inquiry where groups are assigned a question to address. Through this, the experiment can be adapted for students at different levels and is hence a flexible teaching resource.

Examples of other green chemistry laboratory experiments for outreach that are easily transferrable include making a plastic from starch *(80)*. In this activity, students make a plastic from potato starch and investigate the effect of adding a "plasticizer" on the properties of the resultant polymer. This can be used as a laboratory experiment to enhance the teaching of polymers/plastics, an introduction to further work on biopolymers/bioplastics and/or as an example of the effects of plasticizers. Specifically, the problems with using oil to make plastics can be highlighted such as that oil is nonrenewable, there is a danger of oil spills, energy is required to drill for oil, lots of heat is required to separate crude oil, and oil-based products take thousands of years to biodegrade. As such, there is a significant demand for bioplastic alternatives. Conveniently, students can begin either with potatoes or with commercially-bought starch. Furthermore, to aid international transferability, cassava or yams can also be used as a feedstock.

An alternative experiment commonly used in the United Kingdom and that has been implemented by instructors in Brazil is making glue from milk *(81)*. Glue can be made from a protein in milk (casein) where, upon souring the milk with vinegar, it is separated into curds and whey through coagulation and precipitation. The curds can be neutralized through addition of a base to produce a glue. This product can subsequently be used for arts and crafts. Further to being consumed in milk and to make adhesives, casein is also used to manufacture binders, protective coatings, plastics (such as knitting needles), fabrics, food additives, and many other products.

9.7 **Green chemistry education outside the laboratory**

In addition to developing transferrable practical experiments to teach green chemistry principles, there are other teaching interventions that instructors can utilize. With the aim for the education to be effective, learners must be engaged with green chemistry at a deep level *(82)*. One way to achieve this is through active learning where learners are actively or experientially participating in the learning process. In doing so, this enables students to engage in higher order cognitive tasks, resulting in a deeper subject understanding *(83)*. There are a variety of intimately-related teaching strategies that can be implemented to facilitate active learning in green chemistry such as project-based learning, inquiry-based learning, experiential learning, contextualized learning, and cooperative learning where use of technology-enhanced learning is incorporated throughout *(84)*.

There have been several examples of enriching laboratory experiments through activity extensions in the classroom. For example, experiments covering green chemistry, energy production, and environmental degradation were supported by a range of pre- and postlaboratory classroom activities *(78)*. In small groups, students adopt a "jigsaw-type" approach where individuals read a different scientific article before discussing as a group how the articles are interconnected in terms of the scientific study and the link to sustainability. In principle, this approach does not have to be linked to a supporting laboratory experiment that students conduct but can be a standalone classroom-based activity. A similar intervention consists of students evaluating the greenness of a variety of literature protocols proposed for a set synthesis *(85)* (using a metrics tool, Green Star *(86)*). Following this evaluation, students determine which protocol is greenest for each stage (reaction, isolation, and purification) before testing their new combined method in the laboratory. In doing so, this challenges students to reconsider that a set protocol must be used within the laboratory. Taking this a step further, Obhi et al. *(87)* have introduced a combined laboratory experiment and class exercise where students compare the sustainability of two industrial amination reactions to form the same target molecule. Students use industrial solvent and reagent guides, complete process mass intensity calculations, and apply the 12 Principles of Green Chemistry to compare and contrast the reactions. As a summary, students recommend a preferred reaction procedure after considering multiple sustainability concepts in their analysis. Finally, there are approaches to embed science writing heuristics (SWH) into green chemistry experiments to enhance the degree of inquiry associated with an investigation *(88)*. In the classroom, students pose questions and provide methods to address the questions in relation to the laboratory exercises, before carrying out appropriate investigations. By challenging students to think more deeply about the subject via SWH, students demonstrated enhanced environmental literacy.

There are of course (and arguably more transferrable) practices for teaching green chemistry outside of the laboratory. For example, inquiry-based learning in the classroom is often facilitated by students examining a case study or real-world context to provide a scenario for problem-solving. While this adds meaning and familiarity to the learning experience, such static scenarios do not reflect real-world problem solving where both the context and scope of the problem can change with time. To represent reality more fully, a dynamic problem-based learning model was devised where students tackle an individualized problem *(89)*. Different groups of students are provided with different scenarios on sustainable development (including greening transportation), different initial data sets and also during the exercise are provided with additional information that may positively or negatively impact the problem. More approaches to facilitate hands-on learning have also been developed. As an example, Hudson et al. devised a method for understanding different mass-based metrics and determining their strengths and weaknesses through modeling reactions using interlocking building blocks *(90)*. This activity helps students to visualize whole molecules, individual atoms, and mass, further allowing students to make the connection between what is converted into product and what is classified as waste or by-product. Associated with this, interlocking building blocks can be used to aid visual understanding of the synthesis of plastics from monomers and discuss end-of-life options for plastics to include polymer recycling (chemical and mechanical), energy recovery, and landfill *(91)*.

9.8 Use of technology-enhanced learning to share and teach green chemistry

9.8.1 Social media

Technology-enhanced learning interventions can be utilized in conjunction with the former laboratory/classroom activities but also within their own right to educate within green chemistry. As an accompaniment to a laboratory experiment, students can be directed to create an animated and instructional video, posted, with permission, on YouTube based on the equipment setup (with safety considerations) and scientific concepts. This has been applied to an acid rain neutralization experiment in the literature though the concept is transferrable *(92)*. Further to enhancing student understanding of green chemistry, this approach has the advantage of developing student communication skills. Indeed, this work associated with laboratory experiments can be extended to empower students to communicate context-based scenarios with green chemistry principles embedded via YouTube for outreach purposes *(93)*. Through such a setup, students can become global educators in their own right.

Social media can be a convenient mode to engage students with green chemistry especially as it is not dependent on location, with the ability to share content from across the world. Beyond YouTube, there are opportunities to use other platforms such as Twitter, where, for example, information can be communicated in real time through field trips, during which participants tweet classmates who remain in the classroom *(94)*. An empirical study evaluating the effects of Twitter on college students communicating with each other and the instructor revealed the platform had a positive impact on both student engagement and attainment *(95)*.

Another social media platform that has been employed to teach and share green chemistry content is Snapchat. Snapchat is a photo messaging app available on both iOS and Android platforms that allows users to share images and videos (with sound) that can be annotated with text or hand-drawn illustrations. Green chemistry content has been shared with a cohort of students through the establishment of a "class account," which students follow and view image and video updates through the "Story" feature *(96)*. This allowed green chemistry content to be contextualized by students in the real world. Through linking content with real-life examples and communicating them in this way, students are likely to be more engaged with green chemistry together with being able to see the relevance and application of taught material in their daily lives *(97)*. Snapchat can also be used to facilitate research-led teaching in green chemistry; and indeed, this can occur at an international level. For example, during a research visit in Brazil, students received live media of green chemistry experiments being designed within the Amazon rainforest, which they were to subsequently conduct as part of their degree program (Fig. 9.2). This has the potential to encourage students to become excited about green chemistry research, providing a large cohort with a bird's eye view into the world of research,

FIGURE 9.2

Representative snap from developing green chemistry experiments in the Amazon rainforest.

which many students may partake in at a later stage in their studies and careers. As part of the study, further to aiding contextualization of green chemistry and sharing research, Snapchat was used to demonstrate key experimental techniques in the laboratory that students will subsequently practise in upcoming practical sessions. There is also the potential for depicting setups/manipulations with a mistake (engineered by the instructor), asking students to identify the error and subsequently posting an update with the improved/correct demonstration. This alone has educational value associated with students analyzing a system and identifying a mistake. To do so, students must have a good understanding of the correct system to perform their assessment. This could potentially be adapted to compare and contrast a nongreen and green laboratory manipulation. Finally, the instructor used Snapchat to give students an insight into their professional life to humanize the teacher as a real person with the ability for facilitators to enhance the rapport with students. Examples could be extended to teachers/instructors being involved in conferences, meetings etc. Despite this study being limited to one cohort of students, this can of course be extended to any user who wishes to download the application and follow the account.

9.8.2 Online platforms

Further to utilizing social media platforms to support teaching green chemistry education internationally, online platforms to deliver content globally have been constructed in recent years (linking in to one of the recommendations from the White Paper at the Gii in Brazil *(35)*). One example of such an output is the online platform developed as part of the CHEM21 (Chemical Manufacturing Methods for the 21st Century Pharmaceutical Industries), which is Europe's largest public-private partnership for the development of manufacturing sustainable pharmaceuticals (http://www.chem21.eu). CHEM21 brings together six pharmaceutical companies, five small and medium-sized enterprises and research groups from ten universities spanning the breadth of Europe. As part of the project, an entire Work Package (WP) was dedicated to "education and training" which was led by the GCCE at the University of York. This WP was designed to result in the production of an educational program to train undergraduate/graduate students and medicinal chemists from the pharmaceutical industry in sustainable chemistry, capitalizing on the input of consortium partners' experience and benefit from the outcomes of the other WPs as part of the CHEM21 project. The online learning platform (http://learning.chem21.eu) comprises of a range of free, shareable, and interactive educational and training materials to promote the uptake of green and sustainable methodologies, with an emphasis on synthesis of pharmaceuticals *(98)*. Key features of the platform are that it is:

- open (no payment or log-in required to access the site)
- flexible (standalone learning modules so users can create a learning journey appropriate to their needs)

- shareable (most of the content is under a Creative Commons license, allowing for materials to be reproduced and reused)
- interactive (learning resources are provided in a variety of formats including text, video, charts and diagrams, interactive tools, multiple choice quizzes and case studies)

A broad range of topics at introductory and advanced levels are covered to include:

- foundation
- guides and metrics
- solvents
- synthetic toolbox
- process design
- life-cycle impacts and the environmental fate of pharmaceuticals

Furthermore, the platform was specifically designed to facilitate blended learning to maintain user engagement. Rapid feedback questions and more detailed exercises requiring deeper comprehension and application to test knowledge and understanding are also provided. For those interested in learning more, embedded references and further reading are included to allow easy access to more information. The flexibility of the platform for both personal use and for teaching purposes provides opportunities to foster independent learning but also for instructors to incorporate cutting-edge research into teaching, cover core concepts, extend learning, or even facilitate "flipped" lectures. The ACS Green Chemistry Institute Pharmaceutical Roundtable has recently adopted the platform in order for the most recent research developments to be incorporated into it, thus ensuring longevity.

Other alternative online platforms include "Reagent Guides" *(99)*, which was created to encourage users to choose "greener" reaction conditions and "Michigan Green Chemistry Clearinghouse" *(100)*, which attempts to enhance green chemistry awareness, innovation, and investment in the State of Michigan through providing dynamic and interactive online information, resources, database, and interactive tools. The Molecular Design Research Network (MoDRN) *(101)* is a comprehensive database of tools for students and educators at the high school and undergraduate levels that have been produced from educators at Yale University *(102)*. This database focuses on green chemistry and safer design of materials with teaching resources available to download for incorporation into biology, chemistry, or environmental science classrooms. Educators at the University of Scranton have produced green chemistry teaching modules that can be downloaded for integration into courses covering areas such as general, organic, inorganic, environmental, polymer, industrial chemistry, biochemistry, and chemical toxicology *(103)*. Notably, this resource is available translated into Spanish and Portuguese, significantly enhancing the accessibility of the platform internationally. Primarily focused on kindergarten through 12th-grade science (but also at university and professional levels), Beyond Benign provides comprehensive resources, curricula, studies, and programs on green chemistry education *(23)*. Further to such online platforms, the emergence of green chemistry massive open online courses (or MOOCs) is also becoming prevalent with examples such as Biobased Products for a Sustainable (Bio)economy *(104)* and Bio-based Economy-Green Chemistry *(105)*. Presently, more "chemistry-based" as opposed to "bio-based" MOOCs in green chemistry would be welcome in the community. Finally, in order to excite and teach students about green chemistry, a recent gamification approach has been adopted to motivate students at the undergraduate and advanced high school levels to consider green chemistry and sustainability issues as they design a hypothetical, chemical product *(106)*. The game is free of charge and encourages students to think like professional chemical designers to develop a chemical product with respect to function and improved human and environmental health, and in doing so develop an appreciation of the 12 Principles of Green Chemistry *(107)*. Such fun, interactive, and engaging games are likely to foster active learning within green chemistry, allowing students to gain a deeper understanding of subject matter.

9.9 The green formula

A new type of non-lab-based resource has been developed by Glenn Hurst and colleagues at the GCCE at the University of York whereby working with students as partners from across the institution has led to the production of a children's book to engage high school students, aged 12 years upward, with green chemistry *(108)*. As part of their undergraduate degree program, students in the Department of Chemistry collaborated with students in the Department of Education to produce "The Green Formula." Illustrations in the children's book were also completed by a student, following winning an institutional-wide book cover art competition. The book aligns well with the school curriculum in the United Kingdom and follows the story of four very different children who must work together to develop their school's entry to the "National Awards for Technology and Science." This research-led book introduces green chemistry concepts through the diaries of the children and together with an accompanying narrative, provides a number of fun, hands-on experiments and activities to try at home/as part of independent study. This project serves as an excellent example of student collaboration across multiple disciplines together with creating a real sense of accomplishment for students who act as the creators. Such an exercise also opens up new assessment opportunities for instructors and is not limited to green chemistry (though this happens to be the subject of this work). The children's book is available to download for free via iTunes *(108)*, promoting accessibility.

9.10 The ACS Global Innovation Imperative revisited

The recommendations of the white paper from the ACS Gii in Brazil were to:

1. develop transferrable green chemistry experiments and activities

2. construct online platforms to distribute transferrable educational materials
3. initiate a network of green chemistry ambassadors by selecting and training highly engaged educators across Brazil
4. embed green chemistry into the national curriculum in an integrated fashion.

It is clear that a multitude of green chemistry experiments are being developed by the global community, especially in the field of polymer science, though to foster international implementation, an increasing focus on utilizing cheap and abundant reagents and equipment should be made. Further to laboratory experiments, non-lab-based activities that facilitate active learning are also being developed. In relation to the second recommendation, a considerable number of online platforms are emerging to communicate green chemistry at multiple levels with audiences comprising of both students and instructors. Some such platforms are even currently available in multiple languages. Other technology-enhanced learning interventions have also been recently employed to stimulate and engage students with green chemistry such as via MOOCs and social media. In addition to classroom-based games as a teaching aid, there is potential for games to be developed either through online platforms or via mobile applications. Given the widespread use of mobile technologies in the developed and developing world, gamified approaches to teaching green chemistry through engaging mobile applications may gain significant traction in the coming years.

Following the ACS Gii meeting, to address the third recommendation of the white paper, a network of instructors was established where a meeting was held in Rio de Janeiro for high school teachers to insert green chemistry into the school curriculum. This initiative was led by Peter Seidl and Frederico Schoene *(109)*. Training in this meeting comprised of introducing educators to the concepts of green chemistry, sharing implementable practical activities and experiments, together with evaluating the chemistry curriculum and how green chemistry may be inserted into it. Advanced topics such as toxicology and biorefining were also shared with the group through lectures from research leaders in the region together with a roundtable discussion as to how green chemistry can be implemented into the school life. The meeting successfully taught instructors about the concepts of green chemistry together with the facilitation of creating new contacts and ideas to implement classroom activities and low-cost laboratory experiments. So that the conversations can continue, a social media group of educators was formed, allowing instructors to exchange ideas and experiences together. Since the meeting, a number of articles have been published sharing how green chemistry has been incorporated into teaching practice in Brazil *(110)*. Further work includes hosting subsequent meetings to expand the network of educators equipped with green chemistry knowledge and appropriate pedagogies together organizing green chemistry activities as part of national events such as the National Week of Science and Technology. Through continuing to fulfill the first three recommendations of the white paper, working not only at a national level but through sharing transferrable practice internationally,

Brazil can move closer to completing the fourth recommendation of integrating green chemistry into the national curriculum.

9.11 The future for international green chemistry education

The ACS Gii in Brazil and work as a result of this provide an excellent template for implementation of green chemistry education at multiple levels in other developed and developing countries. Through continued global collaboration and sharing of transferrable green chemistry resources, integration into curricula is likely to become more realistic. A significant drive for this has come from the establishment of the UN SDGs and current work to teach green chemistry through a systems thinking approach is likely to enhance the incorporation of green chemistry in curricula at multiple levels.

References

1. *United Nations Sustainable Development Goals,* 2015. https://sustainabledevelopment.un.org/?menu=1300.
2. Welton, T. UN Sustainable Development Goals: How Can Sustainable/Green Chemistry Contribute? There Can be More than One Approach. *Curr. Opin. Green Sustain. Chem.* **2018,** *13,* A7−A9.
3. Mahaffy, P. G.; Krief, A.; Hopf, H.; Mehta, G.; Matlin, S. Reorienting Chemistry Education Through Systems Thinking. *Nat. Rev. Chem.* **2018,** *2,* 0126.
4. Mahaffy, P. G.; Brush, E. J.; Haack, J. A.; Ho, F. M. Journal of Chemical Education Call for Papers − Special Issue on Reimagining Chemistry Education: Systems Thinking, and Green and Sustainable Chemistry. *J. Chem. Educ.* **2018,** *95,* 1689−1691.
5. *United Nations Decade of Education for Sustainable Development (2005−2014): International Implementation Scheme;* United Nations Educational and Cultural Organization (UNESCO): Paris, October 2005.
6. Burmeister, M.; Rauch, F.; Eilks, I. Education for Sustainable Development (ESD) and Chemistry Education. *Chem. Educ. Res. Pract.* **2012,** *13,* 59−68.
7. Burmeister, M.; Eilks, I. Using Participatory Action Research to Develop a Course Module on Education for Sustainable Development in Pre-service Chemistry Teacher Evaluation. *Center for Education Policy Studies Journal* **2013,** *3,* 59−78.
8. Zuin, V. G.; Marques, C. A. Green Chemistry Education in Brazil: Contemporary Tendencies and Reflections at Secondary School Level. In *Worldwide Trends in Green Chemistry Education;* Zuin, V. G., Mammino, L., Eds.; Royal Society of Chemistry: Cambridge, UK, 2015; pp 16−26.
9. Doxsee, K. M. Collaborative Development of a High School Green Chemistry Curriculum in Thailand. In *Worldwide Trends in Green Chemistry Education;* Zuin, V. G., Mammino, L., Eds.; Royal Society of Chemistry: Cambridge, UK, 2015; pp 61−75.
10. Gamer, N.; Huwer, J.; Siol, A.; Hempelmann, R.; Eilks, I. On the Development of Nonformal Learning Environments for Secondary School Students Focusing on Sustainability and Green Chemistry. In *Worldwide Trends in Green Chemistry Education;*

Zuin, V. G., Mammino, L., Eds.; Royal Society of Chemistry: Cambridge, UK, 2015; pp 76−92.

11. Andraos, J.; Dicks, A. P. The State of Green Chemistry Instruction at Canadian Universities. In *Worldwide Trends in Green Chemistry Education;* Zuin, V. G., Mammino, L., Eds.; Royal Society of Chemistry: Cambridge, UK, 2015; pp 179−212.

12. Tarasova, N.; Lokteva, E.; Lunin, V. Green Chemistry Education in Russia. In *Worldwide Trends in Green Chemistry Education;* Zuin, V. G., Mammino, L., Eds.; Royal Society of Chemistry: Cambridge, UK, 2015; pp 213−247.

13. Karpudewan, M.; Roth, W.-M.; Ismail, Z. Education in Green Chemistry: Incorporating Green Chemistry Into Chemistry Teaching Methods Courses at the Universiti Sains Malaysia. In *Worldwide Trends in Green Chemistry Education;* Zuin, V. G., Mammino, L., Eds.; Royal Society of Chemistry: Cambridge, UK, 2015; pp 248−265.

14. Seidl, P. R.; Freire, E.; Borschiver, S.; Leite, L. F. Introducing Green Chemistry into Graduate Courses at the Brazilian Green Chemistry School. In *Worldwide Trends in Green Chemistry Education;* Zuin, V. G., Mammino, L., Eds.; Royal Society of Chemistry: Cambridge, UK, 2015; pp 266−277.

15. Luis, S. V.; Altava, B.; Burguete, M. I.; Garcia-Verdugo, E. Educational Efforts in Green and Sustainable Chemistry From the Spanish Network in Sustainable Chemistry. In *Worldwide Trends in Green Chemistry Education;* Zuin, V. G., Mammino, L., Eds.; Royal Society of Chemistry: Cambridge, UK, 2015; pp 278−307.

16. Haack, J. A.; Hutchison, J. E. Green Chemistry Education: 25 Years of Progress and 25 Years Ahead. *Curr. Opin. Green Sustain. Chem.* **2018,** *13,* 123−129.

17. Lokteva, E. How to Motivate Students to use Green Chemistry Approaches in Everyday Research Work: Lomonosov Moscow State University, Russia. *Curr. Opin. Green Sustain. Chem.* **2018,** *13,* 81−85.

18. Karpudewan, M.; Kulandaisamy, Y. Malaysian Teachers' Insights Into Implementing Green Chemistry Experiments in Secondary Schools. *Curr. Opin. Green Sustain. Chem.* **2018,** *13,* 113−117.

19. Wang, M. W.; Li, X.-Y.; He, L.-N. Green Chemistry Education and Activity in China. *Curr. Opin. Green Sustain. Chem.* **2018,** *13,* 123−129.

20. Gacel-Avila, J. The Internationalisation of Higher Education: A Paradigm for Global Citizenry. *J. Stud. Int. Educ.* **2005,** *9,* 121−136.

21. *G2C2 and Green Chemistry Network,* 2019. https://www.rsc.org/Membership/Networking/GCN/.

22. American Chemical Society Green Chemistry Institute. https://www.acs.org/content/acs/en/greenchemistry/about.html.

23. Beyond Benign. https://www.beyondbenign.org.

24. Cannon, A. S.; Levy, I. J. The Green Chemistry Commitment: Transforming Chemistry Education in Higher Education. In *The Promise of Chemical Education: Addressing Our Students' Needs,* Vol. 1193, Daus, K., Rigsby, R., Eds.; American Chemical Society: Washington, DC, 2015; pp 115−125.

25. Network of Early-Career Sustainable Scientists and Engineers. http://www.sustainablescientists.org.

26. Summerton, L.; Hunt, A. J.; Clark, J. H. Green Chemistry for Postgraduates. *Educ. Quím.* **2013,** *24,* 150−155.

27. Clark, J. H.; Jones, L.; Summerton, L. Green Chemistry and Sustainable Industrial Technology − Over 10 years of an MSc Programme. In *Worldwide Trends in Green*

Chemistry Education; Zuin, V. G., Mammino, L., Eds.; Royal Society of Chemistry: Cambridge, UK, 2015; pp 157−178.

28. Academic Programs in Green Chemistry. https://www.acs.org/content/acs/en/greenchemistry/students-educators/academicprograms.html.
29. American Chemical Society Global Innovation Imperative. https://www.acs.org/content/acs/en/global/international/gii.html.
30. *Workshop White Paper for 2nd Gii Water Forum, Colombia,* 2012. https://www.acs.org/content/dam/acsorg/global/international/Gii%20Colombian%20Water%20Forum%202012%20English%20Version.pdf.
31. Report for More Crop per Drop − Raising Water Efficiency.
32. *Workshop White Paper for 2nd International Workshop on Sustainability and Water Quality, India,* 2014. https://www.acs.org/content/dam/acsorg/global/international/gii2014indiafinalwhitepaper.pdf.
33. Singapore; *Workshop White Paper for WITS Water Forum,* 2014. https://www.acs.org/content/dam/acsorg/global/international/acs_gii_singapore_white_paper_2015.pdf.
34. Nigeria; *Workshop White Paper for Proposed Capacity Development for Water Quality Assessment and Management in Nigeria,* 2015. https://www.acs.org/content/dam/acsorg/global/international/resources/gii-2015-final-paper.pdf.
35. Brazil; *Workshop White Paper for Green Chemistry Experiments for Remote Locations,* 2016. https://www.acs.org/content/dam/acsorg/global/international/green-chemistry-in-remote-locations%20-%20gii-2016-white-paper.pdf.
36. Green Chemistry Centre of Excellence, University of York. https://www.york.ac.uk/chemistry/research/green/.
37. Healey, M.; Flint, A.; Harrington, K. Students as Partners: Reflections on a Conceptual Model. *Teaching & Learning Inquiry* **2016,** *4,* 1−13.
38. Dodson, J. R.; Summerton, L.; Hunt, A. J.; Clark, J. H. Green Chemistry Education at the University of York: 15 years of Experience. *Rev. Quim. Ind.* **2014,** *744,* 27−35.
39. Global Harmonized System for Classification and Labelling. https://www.ccohs.ca/oshanswers/chemicals/ghs.html.
40. Candidate List of Substances of Very High Concern. https://echa.europa.eu/candidate-list-table.
41. Registration, Evaluation, Authorisation and Restriction of Chemicals (REACH) Database. https://echa.europa.eu.
42. Hurst, G. A.; Matharu, A. S. GRASPing opportunities for our postgraduate and undergraduate students. *Forum Learning and Teaching Magazine* **2016,** *40,* 24−25.
43. Cernansky, R. Chemistry: Green Refill. *Nature* **2015,** *519,* 379−380.
44. *Undergraduate Professional Education in Chemistry: ACS Guidelines and Evaluation Procedures for Bachelor's Degree Programs,* 2015. https://www.acs.org/content/dam/acsorg/about/governance/committees/training/2015-acs-guidelines-for-bachelors-degree-programs.pdf.
45. Loertscher, T.; Green, D.; Lewis, J. E.; Lin, S.; Minderhout, V. Identification of Threshold Concepts for Biochemistry. *CBE-Life Sci. Educ.* **2014,** *13,* 516−528.
46. Hurst, G. A.; Bella, M.; Salzmann, C. G. The Rheological Properties of Poly(vinyl alcohol) Gels From Rotational Viscometry. *J. Chem. Educ.* **2014,** *92,* 940−945.
47. Casassa, E. Z.; Sarquis, A. M.; Van Dyke, C. H. The Gelation of Polyvinyl Alcohol With Borax: A Novel Class Participation Experiment Involving the Preparation and Properties of a Slime. *J. Chem. Educ.* **1986,** *63,* 57−60.

48. Isokawa, N.; Fueda, K.; Miyagawa, K.; Kanno, K. Demonstration of the Coagulation and Diffusion of Homemade Slime Prepared Under Acidic Conditions Without Borate. *J. Chem. Educ.* **2015,** *92,* 1886−1888.
49. Morris, E. R.; Rees, D. A.; Thom, D.; Boyd, J. Chiroptical and Stoichiometric Evidence of a Specific, Primary Dimerisation Process in Alginate Gelation. *Carbohydr. Res.* **1978,** *66,* 145−154.
50. Li, X.; Liu, T.; Song, K.; Yao, L.; Ge, D.; Bao, C.; Ma, X.; Cui, Z. Culture of Neural Stem Cells in Calcium Alginate Beads. *Biotechnol. Prog.* **2006,** *22,* 1683−1689.
51. Garrett, B.; Matharu, A. S.; Hurst, G. A. Using Greener Gels to Explore Rheology. *J. Chem. Educ.* **2017,** *94,* 500−504.
52. dos Santos, R. M.; Naas, I. A.; Neto, M. M.; Vendrametto, O. An Overview on the Brazilian Orange Juice Production Chain. *Revista Brasileria de Fruticultura* **2013,** *35,* 218−225.
53. Balu, A. M.; Budarin, V.; Shuttleworth, P. S.; Pfaltzgraff, L. A.; Waldron, K.; Luque, R.; Clark, J. H. Valorisation of Orange Peel Residues: Waste to Biochemicals and Nanoporous Materials. *ChemSusChem* **2012,** *5,* 1694−1697.
54. Djilas, S.; Canadanovic-Brunet, J.; Cetkovic, G. By-Products of Fruit Processing as a Source of Phytochemicals. *Chem. Ind. Chem. Eng. Q.* **2009,** *15,* 191−202.
55. Schneiderman, D. K.; Gilmer, C.; Wentzel, M. T.; Martello, M. T.; Kubo, T.; Wissinger, J. E. Sustainable Polymers in the Organic Chemistry Laboratory: Synthesis and Characterization of a Renewable Polymer from δ-Decalactone and L-Lactide. *J. Chem. Educ.* **2014,** *91,* 131−135.
56. Hudson, R.; Glaisher, S.; Bishop, A.; Katz, J. L. From Lobster Shells to Plastic Objects: A Bioplastics Activity. *J. Chem. Educ.* **2015,** *92,* 1882−1885.
57. Fahnhorst, G. W.; Swingen, Z. J.; Schneiderman, D. K.; Blaquiere, C. S.; Wentzel, M. T.; Wissinger, J. E. Synthesis and Study of Sustainable Polymers in the Organic Chemistry Laboratory: An Inquiry-Based Experiment Exploring the Effects of Size and Composition on the Properties on Renewable Block Polymers. In *Green Chemistry Experiments in Undergraduate Laboratories,* Vol. 1233, Fahey, J. T., Maelia, L. E., Eds.; American Chemical Society: Washington, DC, 2016; pp 123−147.
58. Knutson, C. M.; Schneiderman, D. K.; Yu, M.; Javner, C. H.; Distefano, M. D.; Wissinger, J. E. Polymeric Medical Sutures: An Exploration of Polymers and Green Chemistry. *J. Chem. Educ.* **2017,** *94,* 1761−1765.
59. Tamburini, F.; Kelly, T.; Weerapana, E. Paper to Plastics: An Interdisciplinary Summer Outreach Project in Sustainability. *J. Chem. Educ.* **2014,** *91,* 1574−1579.
60. Zhou, H.; Zhan, W.; Wang, L.; Guo, L.; Liu, Y. Making Sustainable Biofuels and Sunscreen From Corncobs to Introduce Students to Integrated Biorefinery Concepts and Techniques. *J. Chem. Educ.* **2018,** *95,* 1376−1380.
61. Samet, C.; Valiyaveettil, S. Fruit and Vegetable Peels as Efficient Renewable Adsorbents for Removal of Pollutants From Water: A Research Experience for General Chemistry Students. *J. Chem. Educ.* **2018,** *95,* 1354−1358.
62. Mackenzie, L. S.; Tyrrell, H.; Thomas, R.; Matharu, A. S.; Clark, J. H.; Hurst, G. A. Valorization of Waste Orange Peel to Produce Shear Thinning Gels. *J. Chem. Educ.* **2019**. Article ASAP.
63. Walkinshaw, M. D.; Arnott, S. Conformations and Interactions of Pectins: II. Models for Junction Zones in Pectinic Acid and Calcium Pectate Gels. *J. Mol. Biol.* **1981,** *153,* 1075−1085.
64. McDonnell, C.; O'Connor, C.; Seery, M. K. Developing Practical Chemistry Skills by Means of Student-driven Problem-Based Learning Mini-Projects. *Chem. Educ. Res. Pract.* **2007,** *8,* 130−139.

65. Flynn, A. B.; Biggs, R. The Development and Implementation of a Problem-Based Learning Format in a Fourth-year Undergraduate Synthetic Organic and Medicinal Chemistry Laboratory Course. *J. Chem. Educ.* **2012,** *89,* 52−57.
66. Schueneman, S. M.; Chen, W. Environmentally Responsive Hydrogels. *J. Chem. Educ.* **2002,** *79,* 860−862.
67. Muzzarelli, R. A. A.; El Mehtedi, M.; Bottegoni, C.; Aquili, A.; Gigante, A. Genipin-crosslinked Chitosan Gels for Tissue Engineering and Regeneration of Cartilage and Bone. *Mar. Drugs* **2015,** *13,* 7314−7338.
68. Hurst, G. A. Green and Smart: Hydrogels to Facilitate Independent Practical Learning. *J. Chem. Educ.* **2017,** *94,* 1766−1771.
69. Muzzarelli, R. A. A. Genipin-Crosslinked Chitosan Hydrogels as Biomedical and Pharmaceutical Aids. *Carbohydr. Polym.* **2009,** *77,* 1−9.
70. Chen, Y.-H.; He, Y.-C.; Yaung, J.-F. Exploring pH-Sensitive Hydrogels Using an Ionic Soft Contact Lens: An Activity Using Common Household Materials. *J. Chem. Educ.* **2014,** *91,* 1671−1674.
71. Sylman, J. L.; Neeves, K. B. An Inquiry-Based Investigation of Controlled Drug-Delivery From Hydrogels: An Experiment for High School Chemistry and Biology. *J. Chem. Educ.* **2013,** *90,* 918−921.
72. Purcell, S. C.; Pande, P.; Lin, Y.; Rivera, E. J.; Paw, U. L.; Smallwood, L. M.; Kerstiens, G. A.; Armstrong, L. B.; Robak, M. T.; Barranger, A. M.; Douskey, M. C. Extraction and Antibacterial Properties of Thyme Leaf Extracts: Authentic Practice of Green Chemistry. *J. Chem. Educ.* **2016,** *93,* 1422−1427.
73. Novaki, L. P.; Costa, R. T.; El Seoud, O. A. Green Chemistry in Action: An Undergraduate Experimental Project on the Quantitative Analysis of Bioethanol Blends With Synthetic Gasoline Using Natural Dyes. *J. Lab. Chem. Educ.* **2015,** *3,* 22−28.
74. Guron, M.; Paul, J. J. Incorporating Sustainability and Life Cycle Assessment Into First-Year Inorganic Chemistry Major Laboratories. *J. Chem. Educ.* **2016,** *93,* 639−644.
75. Serafin, M.; Priest, O. P. Identifying Passerini Products Using a Green, Guided-Inquiry, Collaborative Approach Combined With Spectroscopic Lab Techniques. *J. Chem. Educ.* **2015,** *92,* 579−581.
76. Fennie, M. W.; Roth, J. M. Comparing Amide-Forming Reactions Using Green Chemistry Metrics in an Undergraduate Organic Laboratory. *J. Chem. Educ.* **2016,** *93,* 1788−1793.
77. Lorenzini, R. G.; Lewis, M. S.; Montclare, J. K. College-mentored Polymer/Materials Science Modules for Middle and High School Students. *J. Chem. Educ.* **2011,** *88,* 1105−1108.
78. Aubrecht, K. B.; Padwa, L.; Shen, X.; Bazargan, G. Development and Implementation of a Series of Laboratory Field Trips for Advanced High School Students to Connect Chemistry to Sustainability. *J. Chem. Educ.* **2015,** *92,* 631−637.
79. McIlrath, S. P.; Robertson, N. J.; Kuchta, R. J. Bustin' Bunnies: An Adaptable Inquiry-Based Approach Introducing Molecular Weight and Polymer Properties. *J. Chem. Educ.* **2012,** *89,* 928−932.
80. Royal Society of Chemistry: Making a Plastic From Potato Starch. http://www.rsc.org/learn-chemistry/resource/res00001741/making-plastic-from-potato-starch?cmpid=CMP00005255.
81. Royal Society of Chemistry: Developing a Glue. http://www.rsc.org/learn-chemistry/resource/res00000459/developing-a-glue?cmpid=CMP00005014.
82. Warburton, K. Deep Learning and Education for Sustainability. *Int. J. Sustain. High. Educ.* **2003,** *4,* 44−56.

83. Freeman, S.; Eddy, S. L.; McDonough, M.; Smith, M. K.; Okoroafor, N.; Jordt, H.; Wenderoth, M. P. Active Learning Increases Student Performance in Science, Engineering and Mathematics. *Proc. Natl. Acad. Sci. Unit. States Am.* **2014,** *111,* 8410−8415.

84. Summerton, L.; Hurst, G. A.; Clark, J. H. Facilitating Active Learning Within Green Chemistry. *Curr. Opin. Green Sustain. Chem.* **2018,** *13,* 56−60.

85. Duarte, R. C. C.; Ribeiro, M. G. T. C.; Machado, A. A. S. C. Using Green Star Metrics to Optimize the Greenness of Literature Protocols for Synthesis. *J. Chem. Educ.* **2015,** *92,* 1024−1034.

86. Ribeiro, M. G. T. C.; Costa, D. A.; Machado, A. A. S. C. 'Green Star': A Holistic Green Chemistry Metric for Evaluation of Teaching Laboratory Experiments. *Green Chem. Lett. Rev.* **2009,** *3,* 149−159.

87. Obhi, N. K.; Mallov, I.; Borduas-Dedekind, N.; Rousseaux, S. A. L.; Dicks, A. P. Comparing Industrial Amination Reactions in a Combined Class and Laboratory Green Chemistry Assignment. *J. Chem. Educ.* **2019,** *96,* 93−99.

88. Shamuganathan, S.; Karpudewan, M. Science Writing Heuristics Embedded in Green Chemistry: A Tool to Nurture Environmental Literacy Among Pre-University Students. *Chem. Educ. Res. Pract.* **2017,** *18,* 386−396.

89. Overton, T. L.; Randles, C. A. Beyond Problem-Based Learning: Using Dynamic PBL in Chemistry. *Chem. Educ. Res. Pract.* **2015,** *16,* 251−259.

90. Hudson, R.; Leaman, D.; Kawamura, K. E.; Esdale, K. N.; Glaisher, S.; Bishop, A.; Katz, J. L. Exploring Green Chemistry Metrics With Interlocking Building Block Molecular Models. *J. Chem. Educ.* **2016,** *93,* 691−694.

91. Enthaler, S. Illustrating Plastic Production and End-of-life Plastic Treatment With Interlocking Building Blocks. *J. Chem. Educ.* **2017,** *94,* 1746−1751.

92. Rand, D.; Yennie, C. J.; Lynch, P.; Lowry, G.; Budarz, J.; Zhu, W.; Wang, L.-Q. Development and Implementation of a Simple, Engaging Acid Rain Neutralization Experiment and Corresponding Animated Instructional Video for Introductory Chemistry Students. *J. Chem. Educ.* **2016,** *93,* 722−728.

93. Smith, D. K. iTube, YouTube, WeTube: Social Media Videos in Chemistry Education and Outreach. *J. Chem. Educ.* **2014,** *91,* 1594−1599.

94. Kassens-Noor, E. Twitter as a Teaching Practice to Enhance Active and Informal Learning in Higher Education: The Case of Sustainable Tweets. *Act. Learn. High. Educ.* **2012,** *13,* 9−21.

95. Junco, R.; Heiberger, G.; Loken, E. The Effect of Twitter on College Student Engagement and Grades. *J. Comput. Assist. Learn.* **2011,** *27,* 119−132.

96. Hurst, G. A. Utilizing Snapchat to Facilitate Engagement With and Contextualization of Undergraduate Chemistry. *J. Chem. Educ.* **2018,** *95,* 1875−1880.

97. Azzarito, L.; Ennis, C. D. A Sense of Connection: Towards Social Constructivist Physical Education. *Sport Educ. Soc.* **2003,** *8,* 179−198.

98. Summerton, L.; Taylor, R. J.; Clark, J. H. Promoting the Uptake of Green and Sustainable Methodologies in Pharmaceutical Synthesis: CHEM21 Education and Training Initiatives. *Sustain. Chem. Pharm.* **2016,** *4,* 67−76.

99. Reagent Guides. https://reagents.acsgcipr.org.

100. Michigan Green Chemistry Clearinghouse. https://www.migreenchemistry.org.

101. Molecular Design Research Network. https://modrn.yale.edu.

102. Centre for Green Chemistry & Green Engineering at Yale. https://greenchemistry.yale.edu.

103. University of Scranton: Greening Across the Chemistry Curriculum.http://www.scranton.edu/faculty/cannm/green-chemistry/english/drefusmodules.shtml.

104. Biobased Products for a Sustainable (Bio)economy. https://www.edx.org/course/bio-based-products-for-a-sustainable-bioeconomy.

105. B. Economy: Green Chemistry. https://www.canvas.net/browse/centreofexpertise/courses/biobased-economy.

106. The Safer Chemical Design Game. https://gwiz.yale.edu.

107. Mellor, K. E.; Coish, P.; Brooks, B. W.; Gallagher, E. P.; Mills, M.; Kavanagh, T. J.; Simcox, N.; Lasker, G. A.; Botta, D.; Voutchkova-Kostal, A.; Kostal, J.; Mullins, M. L.; Nesmith, S. M.; Corrales, J.; Kristofco, L.; Saari, G.; Baylor Steele, W.; Melnikov, M. L.; Zimmerman, J. B.; Anastas, P. T. The Safer Chemical Design Game. Gamification of Green Chemistry and Safer Chemical Design Concepts for High School and Undergraduate Students. *Green Chem. Lett. Rev.* **2018,** *11,* 103–110.

108. The Green Formula. https://itunes.apple.com/gb/book/the-green-formula/id1372738926?mt=11.

109. Schoene, F. A. P.; Seidl, P. R.; Marciniak, A.; Gomes, L. C. A.; Furtardo, L. B. Creation of a Network for the Insertion of Green Chemistry in the Curriculum of the Various Modalities of Education of State of Rio de Janeiro. In *25th Biennial Conference on Chemical Education;* University of Notre Dame: IN, 2018.

110. Application of the Conceptions of Green Chemistry in an Experimental Discipline under the CTSA Approach. http://www.abq.org.br/simpequi/2017/.

Index

'*Note:* Page numbers followed by "f" indicate figures, "t" indicate tables and "b" indicate boxes.'

Printed in the United States
By Bookmasters